Construction Business Management

Construction Business Management

A Guide to Contracting for Business Success

Nick B. Ganaway

AMSTERDAM • BOSTON • HEIDELBERG • LONDON • NEW YORK • OXFORD
PARIS • SAN DIEGO • SAN FRANCISCO • SINGAPORE • SYDNEY • TOKYO

ELSEVIER

Butterworth-Heinemann is an imprint of Elsevier

Butterworth-Heinemann is an imprint of Elsevier
Linacre House, Jordan Hill, Oxford OX2 8DP, UK
30 Corporate Drive, Suite 400, Burlington, MA 01803, USA

First edition 2006

British Library Cataloging in Publication Data
A catalogue record for this book is available from the British Library

Library of Congress Cataloging in Publication Data
A catalog record for this book is available from the Library of Congress

ISBN–13: 978-0-7506-8108-7
ISBN–10: 0-7506-8108-X

For information on all Butterworth-Heinemann publications visit
our web site at http://books.elsevier.com

Typeset by Integra Software Services Pvt. Ltd, Pondicherry, India
www.integra-india.com

Printed and bound in Great Britain
06 07 08 09 10   10 9 8 7 6 5 4 3 2 1

Working together to grow
libraries in developing countries

www.elsevier.com | www.bookaid.org | www.sabre.org

ELSEVIER   BOOK AID
International   Sabre Foundation

Contents

Preface

What you can learn from this book

Most general contracting firms start small—formed by smart and ambitious construction project managers, executives, tradesmen, and occasionally even students right out of construction training, but as accomplished as they may be at what they've been doing they are not likely prepared to take on the range of responsibilities forced on them in managing the business of construction in its entirety. I believe this is the primary reason for the high four-year failure rate that start-up contractors in the United States face. According to research published by the Office of Employment and Unemployment Statistics, US Bureau of Labor, by Amy E. Knaup, only about 43 percent of US construction firms that started up in the second quarter of 1998 were still in business four years later.

A contractor or someone planning to become one can better these odds by identifying and managing the elements of risk. This book offers that opportunity. It is based on the things I've learned, used, and refined as a commercial general contractor in the course of starting and operating my own construction firm[1] for twenty-five years. It spreads these tools and the reasoning behind them out on the table, makes suggestions for their use, and offers a proven business philosophy—knowledge a contractor can set in place from day one to put his construction business on a level playing field with the best-run companies. The information presented here is born of missteps as well as best steps, and both are instructive in building a business that is profitable, enjoyable, and enduring.

My guiding theme in planning and writing this book has primarily been to make available in one place as much as possible of what I learned the hard way (due to

---

[1] Ganaway Contracting continues after more than thirty years, under new ownership at the time of writing, and has also operated as Ganaway Construction.

not knowing enough in the beginning about running a business despite having an engineering degree and several years' experience on the project-owner side of construction) so that interested readers may minimize the pain and risk that rush to fill the knowledge void. Of course, not all risk can be eliminated in construction or in any field, but that risk certainly can be managed if its elements are identified and understood.

Secondly, this book also makes the case for niche contracting, especially chain stores and other light-commercial construction. Niche contracting, or specialization, is a strategy that allows a contractor to become more knowledgeable in a field, be seen as an insider, perhaps sought after, more profitable, and better satisfied with his place in construction. These chain store characteristics practically beg the innovative general contractor to focus on chain store construction. It is my experience that the bid lists are shorter, profit margins higher, negotiated work more common, and owner–contractor working relationships a lot better than are usually found in the open-bid private or public work in which bid error is often the factor that determines the bid-winning contractor (note that I did not say determines the "successful" contractor).

The business management principles and techniques presented throughout this book apply to light-commercial building contractors, subcontractors, and to owners of any small business, regardless of industry.

Here are some of the specific issues discussed in this book:

- How to know whether you're cut out to own and run your own business
- What you must know and do as the owner of your construction firm
- The clear advantages of specializing within general contracting
- Ways you can target, check-out, land, and retain profitable customers (the lifeblood of your company)
- How to select, hire, and keep golden employees (the heart)
- Terms and conditions to include in your bids and your contracts with owners to reduce the chance of disputes and misunderstandings
- *Commandments* you must follow to best ensure that you will be paid what you are owed, including step-by-step change-order procedures necessary to avoid disputes and non-payment
- The strict do's and don'ts of mechanics' liens

- The What, When, Where, and Why of licensing and registration and the extreme risk you take if you ignore the rules
- Terms detrimental to contractors that are often present in owner-prepared construction agreements
- Subcontract terms and conditions most likely to result in best outcomes
- What must be done administratively before you break ground on a project?
- Proactive selection and use of accountants, lawyers, and insurance agents to steer you through the minefields in their areas of knowledge
- The common, sometimes fatal judgment errors contractors make, often during their most profitable times
- The potentially ruinous pitfalls to avoid in insurance coverage
- Why a strong reading habit is so important to your success?
- The personal philosophy and attitude required for success in construction
- Corporate organization and administrative methods
- Links to useful construction, government, and other resources online
- The supreme importance of the human factor, as seen most clearly in chapters exclusively devoted to describing the contractor's role as owner of his firm, selecting and keeping the right employees, marketing, creating customer loyalty, assessing the required personal characteristics contractors must possess, and selecting the right outside professionals

This book is sprinkled with personal anecdotes wherever they can be used to strengthen a point and with pertinent quotes by recognized leaders.

Whether you're a contractor or a key employee, a subcontractor, student or even a chain store executive or businessman in an entirely different field, I promise that you will find ideas, techniques, and principles you can transfer immediately to your management and leadership toolbox. Adopting even a single one of them will pay dividends now and for the remainder of your career.

One indicator of the potential effectiveness of this book is that many of my former project managers who cut their teeth at my construction firm on the principles described here and in turn helped to refine them, now own and operate their own successful firms—largely patterned after the one they trained under and contributed to for years.

Now I invite you to travel with me through this book. The knowledge and references you gain will return the time you spend here many times over in the years to come. It will also have lasting value as a book of reference and for use in training key employees.

While you can start anywhere in the book that applies to a particular need, you will benefit most from the global effect of the entire book.

This book refers to certain business and government organizations specific to the United States. Appendix 3 describes their counterparts in the United Kingdom, other European countries, and Canada.

To avoid distracting the reader's eye, I have avoided construction such as "he and she" and "him and her" the alteration of "she" and "her" with "he" and "him" and the ungrammatical "they" and "them" as neuter singular pronouns. Rather, I use the nonexclusive "he" and "him" to refer to a person.

Acknowledgments

This book is dedicated to my former employees at Ganaway Construction and Ganaway Contracting, whose loyalty and dedication have always been an inspiration to me. The grit and stick-to-it-iveness many of them demonstrated along the way set a standard to be envied by any employer, and assures their success in whatever they choose to do. Many of them now own and operate their own business. I feel a kinship with them that will never fade.

Below are my friends and associates whose time and goodwill I have imposed on while putting this book together, which I could not have done without them:

Jim Bidgood, my construction attorney at Smith, Currie & Hancock LLP, Atlanta, throughout most of my construction career has kept me out of trouble more times than I can remember, both through what I call "Preventive Law," i.e., a rather ongoing dialogue about how to stay out of trouble in the first place, and by coaching me through sticky situations with an eye toward solving the problem rather than agitating it, always seeming to know when to hold and when better to fold. His support during my writing of this book and his valuable feedback on the manuscript have been a source of encouragement for me.

Mark Collins met with me in the content-planning stages and offered his perspective as an instructor in the Building Construction Program at Georgia Institute of Technology and as a construction professional. His input was an influence on the scope of this book. Formerly a successful project manager in my company, Mark now brings together contractors and building owners through his firm Collins RSS in Atlanta.

Khalid Siddiqi, Ph.D., construction department chairman at Southern Polytechnic State University, Marietta, GA, gave me valuable insight into the needs of college and university construction programs and prompted key changes to the structure and content of the book.

Jeff Skorich, partner in Progressive Development Services, Inc., Roswell, GA, reviewed my preliminary ideas for this book's contents and later took the manuscript apart word by word. During the time when Jeff was one of my project managers he was known as our in-house construction encyclopedia. Not only did he read incessantly (and still does), his jobs hummed. Jeff's input on the areas covered in this book greatly enhances its value to readers.

Dancy Stroman, my outside certified public accountant for many years, has celebrated with me on the best days of my business and agonized with me over those that were not so good. Her accounting knowledge, practical advice, and personal friendship were invaluable to me in my business career and continue to be so. She has once again played a crucial role by critiquing my book. I could not have submitted this book to my publisher without the changes prompted by Dancy's suggestions and comments.

Jim Williams, owner of Risc Associates, Atlanta, began his insurance agency at about the time I was starting out and provided my insurance and bonding requirements for the next tewnty-five years. Jim's big-picture approach to my needs and his commitment to detail permitted me to carry out construction projects without concern about unidentified exposure. Now, as then, Jim's review and suggestions for improvement of the chapter on insurance and bonds gave me assurance that this information can be put to use with confidence.

As regional director for a national restaurant chain, Don C. Mathis gave me the opportunity to earn his firm's work before I had much of a track record, in the early days of my construction business. This was the beginning of hundreds of projects I built for Don over several years, which entrenched me in the chain store construction niche. I am eternally indebted to Don for the opportunities he gave me and for his continuing friendship.

I am appreciative of literary agent M. D. Morris of Ithaca, New York, for his valuable advice and for presenting my book to my publisher, Reed Elsevier. At Reed Elsevier, I have had nothing less than the finest relationship with Senior Commissioning Editor Alex Hollingsworth, Editorial Assistant Lanh Te, and Project Manager Charlotte Dawes and her associates, all of whom have made the everyday work of getting a book ready to be published a much greater pleasure than it might otherwise have been.

Writing, like running a construction firm or any other meaningful pursuit, can command the soul at times and eat up all its energy. Throughout it all, my wife, Lee, has provided endless support, encouragement, and enthusiasm, and always a spirit of "What can I do to help?" Having Lee as my wife adds meaning to any successes large or small.

Author disclaimer

This book is a basic guide only, intended to provide the reader with general information about construction business management and the topics discussed herein. The information, opinions, sample forms, and other materials provided are based on the author's personal experience and are provided for general information purposes only, and are not, and should not be, considered or relied upon as legal, tax, or other professional advice. This book is not meant to be a replacement for professional advice and the reader should always seek the advice of a qualified professional in the specific area of need. Neither the author nor any representatives of the author will be liable for damages arising out of or in connection with the use of this book and any of the sample forms and other materials provided herein. This is a comprehensive limitation of liability that applies to all damages of any kind, including, without limitation, compensatory, direct, indirect or consequential damages, loss of data, income or profit, loss of or damage to property, and claims of third parties.

Chapter 1

Do you have what it takes?

Leadership consists not in degrees of technique but in traits of character: It requires moral rather than athletic or intellectual effort...

– Lewis H. Lapham, *Money and Class in America*

Before going further, let's look at some of the characteristics you must have in order to succeed as your own boss. Starting and running your own business are two of the toughest jobs you or anyone can have. You may not have a boss looking over your shoulder telling you what to do, but you actually have the most demanding boss of all. This tyrant lives right inside you, and will demand your attention every minute of your day. He'll be there to ask you hard questions, judge your performance, and nag you about tomorrow's meeting when you go to bed at night. He expects you to risk a lot of your assets, maybe all of them, especially in the early years. It's low pay or no pay for you, but he won't hear to your missing payday for your employees. Most work weeks for you are six or seven days long, and this ogre wants you to be the first one in the office and the last one to leave. He turns thumbs down on the new truck you've decided you need. This boss expects you to do many things at one time and perform them with excellence, even though you haven't mastered all of them yet.

Why would you work for this guy? You ask. The answer is incomprehensible to those who have opted for a more mainstream existence, yet reasonable to us who go out on our own: It's because getting our own business off the ground, creating our own opportunities, blazing our way through the thickets we encounter, and driving forward when the darkest clouds loom give us a confidence, a pride, a joy of life that cannot be bought, bargained for, or willed to us by a rich uncle. It's something that wouldn't be the same if we hadn't paid for it with sweat, tears, worry, and sacrifice. But as we grow and learn, this live-in drill sergeant of a boss begins to give us some slack, some tastes of the milk and honey.

If self-employment as a building contractor is not for the timid of spirit, neither is it for the weak of character. If we lie, we will not be believed. If we don't follow through with our commitments, no one will place their faith in us. If we cheat, we'll be found out—even if not by others, we still know about it ourselves. Aside from any moral argument, lacking character dooms us to mediocrity at best in whatever we do. A sound ethical compass will be at the top of almost any top-ten list of must-have leadership qualities.

1.1 Essential traits

Consider the following characteristics when weighing whether owning your own construction company or other business is right for you. Having even all of them does not promise success, nor does failing some of them now mean you won't rise to meet them when the time comes.

1.1.1 Initiative

You know it if you're capable of managing yourself: Something inside drives you to understand and act on the requirements of your endeavors. You're ever alert for some fresh idea that you can exploit, and you've long cultivated your mind to receive it. You're going over the day's To-Do list before the alarm clock goes off in the morning and you can't wait to get started on it. Each time the phone rings you wonder what new opportunity has found its way to you. You drive home in the evenings thinking of the day's victories, however large or small, and tomorrow's opportunities. You learn from your setbacks and quickly put them behind you.

1.1.2 Passion

Nothing difficult is accomplished without passion. Your passion and intensity not only fuel your own fires but lead others to play above their heads, to do more than is required of them. Winners are passionate.

Jack Welch, the former chairman of General Electric (GE), wrote this about passion in his book, *Jack: Straight From the Gut,*

Making initiatives successful is all about focus and passionate commitment.

Welch also said,

For me, intensity covers a lot of sins. If there's one characteristic all winners share, it's that they care more than anyone else. No detail is too small to sweat or too large to dream. Over the years, I've always looked for this characteristic in the leaders we selected. It doesn't mean loud or flamboyant. It's something that comes from deep inside.

1.1.3 Stress tolerance

Few owners of even the best-run firms in almost any industry have dodged the stresses of dealing with unreasonable characters, costly mistakes, slow-paying customers, employee problems, endless governmental regulation, disappointing financial reports, unfavorable market conditions, angry customers, uncooperative bankers, rising costs, and other challenging situations too numerous to name. To succeed in your business, you must put these in the perspective of your larger picture and move forward. You cannot allow the resulting emotions to dominate your time and energy, impair your decision-making ability, or dim your overall attitude.

1.1.4 Reliability (follow-through)

Do What You Say You Will Do When You Say You Will Do It. Nothing builds credibility better, and nothing is more important to cementing effective relationships with your employees, customers, outside contractors, banks, insurance carrier, and others you depend on for your success. Keeping commitments is your Eleventh Commandment. One-time neglect to return a phone call can leave a lasting memory. Failure to keep a major commitment such as a delivery schedule can fracture the good image you've spent years building.

1.1.5   Willingness to work while others play

Owning and operating a construction firm is not a five-day, eight-to-five occupation. You can count on six or even seven long days in the beginning while you're setting up your company's infrastructure, simultaneously building projects, and trying to get new ones. As you become more established, plan on fifty- to sixty-hour weeks routinely, and often longer. Your vacation, when you get one, will no longer be a time when you can forget about everything back at the office. Not all of these hours will be at the office or jobsite. Some will be spent at home in the evenings or on weekends planning, reading, or just mulling over business matters.

If you are unable or unwilling to give this amount of time to your business, reconsider starting your own business. However, for most self-employed business men and women, the hours aren't burdensome. They fly by, and the day is gone before they want it to be.

1.1.6   Unyielding positive attitude

Thomas Jefferson said, "Nothing can stop the man with the right mental attitude from achieving his goal; nothing on earth can help the man with the wrong mental attitude." The "right" mental attitude is belief in yourself that you can and will defeat the obstacles that come your way and turn them into opportunities. The writer Charles Swindoll says, "... life is 10 percent what happens to me and 90 percent how I react to it."

1.1.7   Mental toughness

There are times when it gets lonely for the sole owner of a business. You're forced to make decisions that you know will not have perfect outcomes. Sometimes they are agonizing, yet there is really no one who can totally comprehend what you may be going through because it is impossible for another person to relate to the unique set of financial and practical consequences you face. That's when you draw on your belief in yourself; remember other rough times that you've pulled through; and remind yourself that you did all your homework, dotted all the i's, and never expected it to be all roses. The ancient philosopher Epictetus said, "Difficulties show men what they are."

1.1.8 Attention to detail

The attention you give to detail becomes evident in most aspects of your life, such as your personal appearance, your home, even the condition of the car or truck you drive. In your construction business, your attention to detail shows up in the people you hire, the bids you send out, the quality of your work, and the image presented by your organization to prospective customers and countless others whose opinion may have an impact on your success.

1.1.9 Sense of urgency

In many situations there is a brief moment during which an opportunity must be grasped or it is lost forever.

1.1.10 Self-control

Being on time. Keeping a cool head in the face of adversity. Spending within your means. Practicing sensible personal habits. The list goes on. These impact both your personal and your business life.

1.1.11 Thirst for knowledge

Knowledge of construction and of running a business are only the beginning of what you need to know. Politics, the stock market, interest rates, general business trends, who's who and what's what in your customers' industries, and knowledge of your competitors are also among the things that require pipelines of information flowing to you. Reading newspapers, magazines, and books that provide such information will put you a step ahead of much of your competition, and so must become part of your daily routine.

1.1.12 Ability to get along with others

There may be no more valuable skill than the ability to deal successfully with people. The human quest for this skill is demonstrated, for instance, by the enduring demand for Dale Carnegie's book, *How to Win Friends and Influence People*. Based on his extensive research, Carnegie declared that financial success is due 15 percent to professional knowledge and 85 percent to "the ability to express ideas, to assume leadership, and to arouse enthusiasm among people."

Again, don't be discouraged if you haven't yet conquered all of these traits. Internalizing their importance gets you halfway there. Working with purpose and determination can take you the rest of the way.

Most of us who've taken the plunge into entrepreneurship wouldn't go back if we could. Owning our own business is not a destination, but a journey we hope never ends because we have the best seat on the train, one from which our senses are brimmed over by spectacular mountain peaks, rushing whitewaters, and lush mountain valleys. In America and other free-market nations we have the freedom to excel as well as to fall flat, and we learn from both. Many entrepreneurs thriving today suffered initial failures, then rebounded with new wisdom, cleared away the debris, and started all over again.

Chapter 2

Your role as owner of your construction firm

Remember, you build a business from the top down and from the inside out. Be assured you lead, not push, your way to profitability. This holds whether it is a two-person firm or a 200,000 corporate behemoth...

– Michael H. Mescon and Timothy S. Mescon, *Atlanta Business Chronicle*

As owner of your own business, you assume three major titles: Leader, Manager, and Stockholder. Until you fulfill the responsibilities implied in these titles, they mean little more than the fact that your name is next to them. You will define these roles uniquely for yourself as you grow in each area of responsibility. Below are the challenges you will face.

2.1 Leadership (Setting the course)

A leader is someone who visualizes and communicates an objective with so much passion and influence that people around him internalize it and deal with it as their own. Instead of being driven by their leader, they follow, yet they are the source of the leader's power.

Former top military leader and US Secretary of State Colin Powell says, "... leadership is the art of accomplishing more than the science of management says is possible." And speaker, author, columnist, and psychologist John Rosemond describes leadership in these terms:

> **Leadership is not enabling; it is challenging. Leadership is not indulgent; it is inspiring. Leadership is not permissive; it upholds high expectations. Leadership is not compromising; it is exacting. Leadership is not about the clever manipulation of reward and punishment; rather, it is primarily a matter of how one communicates.**

Backed by your enthusiasm, passion, vision, and credibility, a crisply defined mission in which you specify goals, how you plan for them to be accomplished, what each employee must do, and the rewards at stake can create in your employees an obsession with winning seen in the teams and among fans when such archrivals as the National Football League Dallas Cowboys and Washington Redskins play each other. Your goal becomes the employees' goal, your success their success, and their contribution is likely to far exceed the ordinary. Impassioned leadership can generate a level of enthusiasm beyond the reach of any other motivator.

The "rewards" that motivate your employees may or may not be tangible. Military leaders say soldiers don't fight for money, they fight for a cause. Passionate baseball fans have nothing to gain (assuming no betting is involved!) if their team wins the World Series other than pride in "their" team and bragging rights. Former President John F. Kennedy's goal, presented before the Congress in 1961, of going to the moon within that decade jump-started one of the great national efforts in history to put together the apparatus needed to accomplish his vision. In an article about the Apollo program, David Scott writes that at the peak of the program the National Aeronautics and Space Administration (NASA) employed 36,000 civilians, 376,700 contractor employees, had an annual operating budget of $5,200 million (dollars in 1965), and landed 12 astronauts on the moon—a program that yielded undisputed, wide-ranging, enduring economic and technological benefits. While this is an extreme example it shows the power of vision and leadership, which can be every bit as effective in even the smallest organization. Former President Kennedy's goal became "our" goal, and exemplifies the effective path of all worthwhile visions and ideas.

2.1.1 Vision to reality: The required path

Obviously, action is required if a vision is to become reality. Here is the required route from idea to making it happen.

- **Birth of the vision or idea**   Kennedy envisioned the seemingly preposterous goal of landing Americans on the moon within a given time frame.
- **Go/No-go decision**   Before announcing his goal, Kennedy must have determined to his satisfaction that the goal was doable. Absurd ideas die at this stage.
- **Commitment**   Under Kennedy's initial leadership, Congress and NASA committed themselves to carry out his vision.
- **Plan**   NASA developed plans to accomplish the mission. In this case the plans were necessarily huge and complex. Your plans for carrying out the goals you set for your company will be somewhat more manageable.
- **Execution**   Simply put, NASA made the plan happen. Of course, its difficulty defies characterization.

Americans felt unsurpassed national pride when Neil Armstrong stepped onto the moon in 1969 with those now immortal words, "One small step for man, one giant leap for mankind."

As soon as possible after a mission is achieved make a big deal of it. As former GE Chairman Jack Welch says in his book *Winning*, "celebrate." Recognize the key players as well as the supporters, and distribute the promised rewards with a degree of fanfare. Then identify the next giant you want to slay, and lay out a new plan to make it happen.

Your job as leader doesn't require you to be smarter or know more than everyone else in the company. Certainly, John F. Kennedy did not have the kind of training and knowledge those who carried out the moon mission had. Wise leaders hire people who know more than they do and who complement their own strengths. As the one who has to keep the big picture in focus, you must know your industry, and you have to know enough to ask your key people the right questions, evaluate their responses, redirect them when your instincts tell you it's necessary, and hold them accountable.

The bonds that enable you to lead are based on trust, which you must build and protect with your every word and deed. Once trust is broken, it can be impossible to repair.

2.1.2   Leaders and managers are different from each other

In *On Becoming a Leader*, Warren Bennis differentiates between leaders and managers: "The manager does things right; the leader does the right thing." Leaders have followers. Leaders assume responsibility for whatever happens, they convey a sense of purpose and meaning, they envision the future, and they are contagious.

On the other hand, entrepreneurial leaders don't always have the skills or temperament to be the most effective managers, and often must hire one to carry out their vision. Most intelligent, motivated people can be trained as managers—to organize people, money, time, information, and systems and procedures. If after getting your start-up off the ground you decide that you may not be the manager your company needs, then develop a strong manager who can complement your leadership. (Learn about hiring the right people in Chapter 11, "You and your employees.")

Many successful leaders and managers will disagree with my view, but I like to operate as a leader *among* rather than *of* my people. As your organization's successes grow, some of your employees will see you as more important than themselves, bigger than life even, as we all want to think our leaders are better than us. But shake off the tendency to accept the glory, remembering that a leader is nothing without followers. Know who you are and resist the occasional temptation to portray a different you. Your employees will appreciate your realness, and your power to lead will only grow from it.

2.1.3   Tame the ego

A certain amount of ego is inevitable in a high-achieving person, who probably would not have excelled without a lot of self-confidence. But you're not likely to meet a successful business owner who does not attribute some measure of his success to good timing or good luck or some combination of the two. Certainly, intelligence, leadership, hard work, and other qualities are involved but there's not much room for ego in the equation.

Once when I was a junior-level manager with Shell Oil Company not long out of college, my boss praised the outcome of a project I was responsible for in the presence of a valuable, long-time outside vendor and me. It was the vendor who had actually done the job I had given him, but I hadn't shared the credit with him at that moment when we were all there together. My boss privately pointed out my error to me later. It was a lesson in humility that I haven't forgotten in all these years. Pass the hurrahs on. This principle is especially applicable to your relationship with your employees, e.g., when a customer makes a positive comment about some aspect of a project. Give those responsible the credit they deserve.

2.2 Leadership in times of uncertainty

There's no circumstance in which leadership is more critical than in times of crisis or doubt. Author and speaker John Maxwell talks about the characteristics of leaders and their followers in times of uncertainty. He says leaders study the actions of earlier successful leaders, give hope to their followers, convey compassion, act with courage that others can draw on, and provide a feeling of security to their people by staying close to them. In the *Atlanta Business Chronicle* Maxwell quotes from *Time Magazine's* 2001 Person of the Year story about New York Mayor Rudy Giuliani to show that the mayor provided such leadership not only to New Yorkers but to the nation and parts of the world in the aftermath of the terrorist events of 11 September 2001.

Maxwell says such behavior on a leader's part is necessary to meet the needs of his followers, who in tentative times are searching for security, and hope, and direction.

While businesses often focus on challenge coming from outside the organization, crisis often stems from within, as was the case in the following example in my history. When my company was about ten years old we had developed several new chain store customers during the past year and our "old" customers were going stronger than ever. We had ended a good year, and had built up some cash reserves over the longer term. All of this put me in an expansive mood. So, over a period of months I hired a couple more project managers, increased the administrative and accounting staffs, extensively expanded and remodeled offices, created a second layer of management between me and my jobsite superintendents, and implemented the latest new construction software and hardware (which was very expensive in those early days of computers and specialized software). All of this meant increased overhead, confusion and loss of time learning and converting to the new system, the dead-cost transition period for new employees, and a focus on infrastructure rather than operations.

At this point, I had made two unforgiving mistakes: First, because of the good results of that prior year, and a good history altogether, I lapsed into the faulty thinking that the next year and the next and the next would be just as good. Secondly, I had taken my eye off the ball—the business of running my business—and with all the extraneous activity I had put into motion, I had set up my employees to take theirs off the ball as well. All that non-productive in-house activity was time-consuming and distracting to all of my employees and to me, and it was eating cash voraciously. Even without the events that were to follow, described below, the consequences of my management errors were sufficient to, at best, throw any small business off course.

Within the next few months, three of my customers who had shown every outward indication of continuing business as usual filed for bankruptcy. The economic climate was bad for other of our customers as well, and we had trouble getting paid on several projects we had under construction. We had disastrous performance problems on one of our jobs that ended up with a large loss.

First went our working capital, then our cash reserves. When my audited year-end financial statements were completed, they showed that the company had lost twice the amount of money the in-house monthly statements had been showing. This started the dominos to tumble. The new bank I had switched to before all the trouble began cancelled my line of credit and called in the balance due. My surety company of ten years cancelled my bonding line. I was faced with going beyond the company's reserves well into my personal assets. This was not an easy decision. Without operating cash, bonding availability, and credit line, and with slow- or non-paying customers, business failure was a real possibility. However, a cash infusion from my personal accounts was no promise of success, and failure then would mean the additional loss of the money I had loaned the company, and a blow to the wherewithal I would need to start over in that event.

I had lost control by failing to lead and manage well. Now there was a crisis and leadership was doubly required. I put the money in, and with this the company had digested new cash, including its reserves, amounting to a million dollars just to see another day.

My job now was to lead the company to safe ground. There was a strongly positive side to the equation: I had the goodwill and dedication of my employees and a long record of good business practices with our customers, suppliers and subcontractors, insurance and bonding agent, and certified public accountant, and I drew heavily on these precious intangible assets. We contacted our materials vendors with an explanation and a request for more lenient terms: Sixty to seventy-five days to pay the accounts that we had scrupulously paid in thirty days historically, which would result in a significant improvement in our cash position. I don't recall that even one of the suppliers balked. Most of our small subcontractors required more frequent payment but bent as much as they could. My project managers and in-house accountant met weekly to decide how to divide the still scarce cash among the creditors. I approached a bank executive at my former bank who agreed against all of the bank's policy guidelines to issue an irrevocable letter of credit for my use in securing a bonding line. With this, my bond agent was able to arrange for the performance and payment bonds I needed. Some of my key employees agreed to take cuts in pay, and I suspended my own.

From the beginning I offered encouragement to my employees. I let them know I understood what they were going through and that I was going through it with them. I slashed overhead to the point of agony. At the same time I needed more than ever to keep business rolling in the door and find ways to further reduce overhead. My managers' jobs were to carry out their usual functions plus bear the burden imposed on them by our financial plight. They were used to a lot of autonomy but now they needed a reason to believe that what they were going through was not futile. I agonized with them but failure was never discussed as an option. There were frank discussions about the hill we had to climb but no whining and no what-ifs.

My employees deserve the credit for our survival and I owe them a debt of gratitude that can never be fully paid. It took us two years to recover the losses and get back to "normal" operation, and to my knowledge none of our customers were ever negatively affected by our financial problems. All of our creditors were paid what they were due, even if late.

I take full responsibility for the problems of that time in the mid-1980s. It is true that there were some circumstances out of my control, such as the bankruptcies, but part of my job as owner of my firm was to anticipate what may go wrong and to have a plan in place to deal with it. It's not practical to plan for an asteroid striking Earth or a nuclear attack, but bankrupt and slow-paying customers and vicious economic cycles must be included in your business planning.

What did I learn from this experience? That being laser-focused on the part of my business that generates income requires the highest priority. That past performance is no promise of success going forward. That there is no substitute for dedicated, capable employees. That I need to stay as close as possible to where the money is made or lost with the least possible insulation separating me from that. That building up cash reserves is essential. That managing relationships fairly with business associates during good times builds up a cache of goodwill and trust that can be drawn on in not-so-good times. That calculated risk is necessary for success. That contingency plans must be in place for all practically imaginable pitfalls. And that without effective leadership, ill winds are certain.

2.3   Manager vs owner/shareholder

If your company is like most small construction firms, you have the twin but separate responsibilities of owner and manager. As owner, you decide the goals and objectives for the business. As manager your job is to carry them out by organizing, planning, controlling, directing, and communicating.

Several times each year, you must stand back from your desk and objectively compare your performance as manager against the objectives you previously set as owner. You may be satisfied with the results you see or you may find them not what you had expected. If the latter is the case, you as manager must dig out and "report" the reasons for your less-than-expected performance. Then you the owner must decide what internal changes are required that you the manager must now accomplish.

The responsibilities inherent in your split role as owner and manager are some of the most demanding you'll have to impose on yourself. Failure to separate and fulfill and overview each role independently can result in acceptance of things the way they are, instead of recognizing any departure of performance from expectations and self-imposing the discipline necessary for adjustment. The problem is that things don't remain "the way they are": Without critical evaluation and effective management, they go downhill.

My bet would be that the marginal or failing performance of many privately held firms could be traced to the neglect of the owner to acknowledge and act on the dual responsibilities of owner and manager. What is not looked at with a discerning eye probably will not reveal its opportunities for improvement until it is too late.

2.4   The entrepreneur mindset

People who envision and create their own business usually do so on the entrepreneurial model, in which they keep the strategic planning and problem solving to themselves. Having had the dream, taken the risks, and birthed their business pretty much by their own courage, wit, and determination, they don't easily share the controls with others—no matter how capable their employees or advisors may be. For many entrepreneurs it's much easier to direct others than to assign areas of authority and responsibility, require accountability, and allow capable people to operate with autonomy.

Certainly many businesses thrive for years and years in the entrepreneurial mode, but with sacrifices. Considering various points of view is almost always better than making decisions in the vacuum of The Boss's Office in which the knowledge, experience, and capabilities of your employees and outside advisors are waived off without much thought. Worse, the business will struggle if you're put on the sidelines for health or other reasons before you shake the entrepreneurial mindset; even though the remaining employees may be excellent in carrying out their assigned duties, they will not be accustomed to making decisions that affect the company as a whole, may not know all of your outside advisors, and may not learn how to control the monster quickly enough.

That is not to say you should not be the final authority in all matters. It needs to be clear that once you've considered others' input—as you'd better—it is you who makes the final decision. If your decision at times seems to fly in the face of all the advice you've received, that doesn't mean you could have or would have made the same decision without it.

In weighing possible courses of action, decide *against* other avenues as much as you decide *for* the one you choose.

2.5   Managing risk

Let's face it: Construction is a risky business. But no business worth spending your time in, that I'm aware of, is risk free. The first step in managing risk is knowing where it lies. Some areas of contractor risk are obvious: Estimators make mistakes. Contractors' payment applications are delayed. Owners don't pay on time. Contractors cannot or do not reduce fixed overhead fast enough in rough times. It rains for two months while the job is coming out of the ground. Subcontractors don't perform as required, or go broke. Suppliers don't meet schedules. The cost of certain materials spikes unexpectedly. Jobsite accidents occur. The insurance company denies a claim. Owners dispute claims for change orders. A key employee quits. Non-documented site conditions present themselves mid-project. Owners file for bankruptcy. Contractors' cash runs short. Suppliers cut off credit. And on and on.

We take for granted the ready availability of common building materials, but many essential items including steel, concrete, plywood, and gypsum board have become scarce from time to time. Custom materials made for a specific application are potentially more of a problem, and it's up to you to evaluate the reliability of their chains of supply, whose failure to deliver could severely hurt your job. You may be able to identify a backup source in advance in the event of a primary vendor's failure to perform, but even that solution may increase your project cost and time.

If the lawyers, accountants, insurance agents, and other professionals you rely on focus on the construction industry, and more particularly within your niche of the industry, these experts can help you identify your risks so that you can deal with them effectively.

Your construction-oriented certified public accountant (CPA) can provide financial ratio guidelines based on published statistics as well as his experience with other contractors. Your construction attorney can tell you the top five legal problems his contractor clients have, and help you avoid them. Your insurance agent whose clients are primarily contractors can identify your exposures as a contractor and steer you toward the necessary coverages. Being without this kind of input from people who specialize in your business can leave you in the position of *not knowing what you don't know*. You can deal with a situation once you know it exists.

The bottom line about risk is that you must identify your various exposures, decide the potential impact of each, and do what you can do to manage it both internally and externally. For those exposures that cannot be eliminated, you must develop plans that can be implemented quickly if required to minimize the impact on your company. Managing risk is a persistent theme throughout this book.

2.6 Establishing your corporate culture

A company's culture might best be understood in the answer to the question, "What's it like to work here?" For instance, can you tell the boss when you think he is wrong? What employee behaviors get rewarded? Ignored? Punished? What's really important around here, anyway?

Every company has its own culture whether by design or by default. If by default, the owner and the employees may not even be conscious of their organization's culture, but it exists just the same. You may never write a memo that states a dress code or proclaims that you're open to new ideas, but any employee that's worth the space he occupies in your office watches your behavior and listens to your words for clues to your priorities.

Michael H. Mescon, founder and chairman of Atlanta, GA-based business consulting firm The Mescon Group (now HA&W Mescon Group), says organizations are built from the top down and from inside out. The quality of the employees you hire, the way you relate to them and to people outside the company, and the systems you establish for responsibility and accountability all communicate your values and expectations to the people who work for you.

There are visible manifestations of your firm's culture, such as the type of clothes your people wear to the office, the cars or trucks they own, and the office décor and furnishings. All are not necessarily good or bad, they just exist.

Less visible but real clues of the culture that exists in your firm might include failure of project managers to keep each other informed about their projects, which limits cross-coverage in case of sickness or other absences. Both visible and invisible signs of corporate culture may operate on the conscious or subconscious level, and may be productive or counterproductive in either case.

Here are other examples of the way culture develops within and to some degree defines an organization. If you average a ten-hour workday yourself, your project managers are not likely to drag in late or leave urgent details hanging when they go home for the day. If you wear pressed khakis to the office regularly, you probably won't see office employees in cutoffs and sneakers even if you've never mentioned such an expectation. If you usually reject your employees' suggestions that involve new challenges, don't expect them to bring in unusual projects. If you convey the policy that you want bills paid on time, you won't see your bookkeeper sitting on past-due vendor invoices. If you're cool-headed in a crisis situation (as you'd better be) your employees are probably going to examine how they might react in similar situations.

Your company's culture influences how your employees think about you, and how they feel about the company and even about themselves. It affects how they respond to emergencies, how they handle themselves on the phone or in meetings, how they manage their personal space, and the way they represent your company to others.

You may have written a set of values that you've hung on the walls around your office that describes how you'd like your company to run. It might include words and phrases like, "We strive for excellent quality," "Customer care is priority one," "Empowerment of our employees...," or "Integrity above all." But these are ideals (and generalizations) and don't always describe the culture that actually exists in the firm.

Unfortunately, your culture is best seen by your subcontractors or your attorney or your customers. *Only after you see the gap that exists between your ideal values that you want to promote and the entrenched culture that actually operates can you tag the work that must be done.* That is, until you synchronize what actually exists with what you want, it will be impossible to follow the course you've charted: You'll try to steer in the direction you want to go, but your ship will go in another.

So, knowing yourself—truly seeing yourself (your firm) as you really are—is the first step in putting your values and your culture in sync with each other. Only then can you mold your company through a combination of your behavior and the express values and beliefs you put on paper. However, you cannot fake it. As a small business owner close to your employees, you are more or less transparent. Don't try to sell beliefs and values that are different from the real you. No one will buy them.

2.7 Striving for excellence

You will see the word "excellence" used many times in this book. This is no accident. While it is possible to fail in any endeavor even with the strongest dedication to excellence, I have rarely seen enduring success in any individual or business organization in which excellence at all levels was not demanded and practiced *from within*. Excellence is not just "getting by" with the minimum requirements of the task or project at hand and it is not merely a slogan. It means that what may pass as "satisfactory" performance to others is unacceptable to you. It is an internal standard that you and your top-echelon employees establish for every level of your organization and demand that it be met—in the reports your bookkeeper produces; for your cost estimates; in placing the concrete foundations you build; in the finishes in the buildings you turn over to project owners; for your relationships within your company and with your customers and others outside; in the image you and your employees project for all the world to see; etc.

If you as the leader of your organization do not practice excellence yourself, you will not be able to successfully demand it from your associates, and you and your customers will not see excellence in your company on any consistent basis. The concept of excellence cannot simply be worn as a badge. It must come from your conviction that it is crucial to your success and from your unrelenting requirement that it be met as it is defined for the various functions of your firm. Your employees, and certainly those in managerial positions, must internalize it as well. If they do not or cannot internalize it, i.e., make excellence their own conviction and not just go through the motions, you will be frustrated in your efforts to grow the kind of organization you want and to provide the level of service and product quality that is essential to lasting success.

Don't confuse excellence with perfection. Set as your goal, excellence is consistently achievable. Perfection is an ideal that is neither practical nor affordable in the commercial world.

2.8   Hiring the right people

One of your critical functions as leader is to identify, hire, and groom your firm's future talent for key positions within the company and, some day, even your own. Of course, in the early stages of your business you may fill all of the key roles yourself.

No matter how small your business is, or how large you grow it, your success will depend a lot on your skills in bringing the right people on board, and how you and they relate to each other. This applies to your employees, subcontractors, the lawyers, accountants, and other professionals you bring into your team, and your customers.

Donald Trump's book *The Way to the Top: The Best Business Advice I Ever Received* is a compilation of business advice from contributors whom Trump calls "many of the brightest, most-successful business people" he knows. It impressed me that so many of them write that their success is largely a result of hiring only the most capable people.

Don't compromise. Before you decide to include someone in your business circle, thoroughly research his qualifications with respect to your requirements, and his business and personal reputation. Don't base decisions on initial cost alone.

A commonsense rule is to hire only people you like. Regardless of an employee's qualifications, you're probably not going to stick with him long term if you don't like him and vice-versa. On the other hand, of course, personal compatibility is no guarantee that he will satisfy your business needs.

Hiring and keeping the right people is discussed in Chapter 11, "You and your employees."

2.9 Knowing your industry

It seems unlikely that anyone would risk his savings, time, and effort on a business he doesn't thoroughly understand but many contractors and other entrepreneurs do. You should know the basics of the construction business in general and of your selected field within the industry; its role in the national economy; and how it's affected by various economic, regulatory, and political factors. You learn these things through experience, by paying close attention to local and national events and trends, and through construction industry advocacy groups. You are gaining some of the knowledge you must have by reading this book, but it is only a start. There's no substitute for business-reading on a regular basis.

Routine reading is part of your continuing education. Your personal databank of local, national, and international events and circumstances serves as a fertile ground in which ideas can take hold—ideas that can spring from being tuned in to what's going on in your marketplace, again, partly from reading. Scanning trade publications, print newspapers, and Internet news sources is like mining for gold: You can always expect to find something of value to you or your business, something you can clip and keep or send to a business associate or friend.

The construction advocacy groups Associated General Contractors of America (AGC) and Associated Builders and Contractors (ABC), and, outside the United States, United Kingdom's Construction Confederation (CC), the Canadian Construction Association (CCA), and the European Construction Industry Federation (FIEC), provide much more than industry advocacy. Through member organizations, each offers seminars, training sessions, reading materials, and other educational opportunities, and membership in them indicates a level of competence and professionalism. Members of their local chapters, though often competitors, get to know each other, become friends, discuss problems that all face from time to time, and hatch fresh ideas for approaching them.

You can get a lot of ideas from organizations that are linked to your field of specialization within the construction industry, e.g., retail stores or shopping centers.

As your company grows, you will sometimes wonder whether you know enough to keep up with it and to grow it to the next stage. The familiar saying "It's lonely at the top" is true partially because it's not possible for another person to imagine your identical circumstances.

The time will come in your growth when you will need an outside advisor or mentor. Your CPA and your attorney will be helpful, and you will depend on their advice in the course of business. But close friends, family members, and people you do business with regularly including those professionals may not give you objective opinions outside their fields, finding it uncomfortable to disagree with your ideas. Seek out two or three small- to medium-size business owners who are successful in their fields and approach them on the subject of meeting you once or from time to time to discuss business issues. Some will and others will not. Choose people whose experience can fill in the gaps in your own. Offer to pay for their time so they will understand that you want more from them than casual conversation.

Having some commonality with a person you're considering as a mentor is necessary to get that ball rolling. Your CPA may suggest someone who might fill the bill and arrange an exploratory meeting for the two of you. Another source is your lawyer. In your search, consider personalities as well as knowledge. The relationship won't get far unless there's some favorable chemistry between you.

Becoming active in civic organizations is another excellent way to meet people who may be just the right advisors for you. This way, you have the advantage of learning some of their characteristics before deciding whether to approach them.

2.10   Coordinating resources

It's up to you to establish and maintain the basic elements required to carry on the business. You may delegate some of these functions to key employees, but others require handling by you, the owner, especially in a small company, and always in the beginning. Those you should handle yourself include controlling your finances; maintaining banking relationships, including working capital and line of credit if necessary; securing business and project insurance coverage; providing office space and business equipment; implementing certain business and field operating procedures including the handling of bank statements, cash receipts, and payment of invoices; coordinating marketing and business development; selecting your outside professionals (CPA, lawyers, insurance agent); and hiring key employees.

Your foremost resource is your employees, and it's solely your responsibility to identify and develop key personnel and future leaders of your company. Don't delegate this critical responsibility.

2.11   Keeping in touch

In a start-up business the entrepreneur is likely to be the person answering the phones and taking out the trash. Whether you're in that stage or have grown beyond, it's up to you to be sure this machine you've created doesn't veer outside your established parameters. You may do this by staying in touch to the appropriate degree with every aspect of your business, even as you grow larger.

This doesn't mean that you micromanage. If you've put the right people in place, that's counterproductive. But if you frequently drop by one of your project managers' office and sit down with him for a few minutes you'll get a sense of how things are going on in his projects. When you stop at a jobsite, look around, and chat with your superintendent and subcontractors, you'll know more. (For instance, a disorganized, unkempt jobsite always raises a caution flag to me.) If something doesn't feel quite right to you, it probably isn't, so delve into it until you know why. Compare what you observe on your own with the reports you get back at the office.

Take a break in your bookkeeper's office. Do you get a sense that the financial paper flow is running smoothly? Can he explain to your satisfaction what's behind the numbers on the reports? Ask probing questions about your financial statements until they make sense to you.

Author and businessman Pat Croce talks about his "Five-Fifteen" management tool. On Fridays, he says, each of his employees spends fifteen minutes writing a progress report on his own goals and successes for the week and emails them up the line to the next supervisory level. Eventually the reports trickle up to Croce, who spends five minutes reading each one and perhaps sending it back down the line with an appreciative comment (thus the "five-fifteen" terminology). He says the writers see it not as a chore but as an opportunity to showcase their achievements. And to Croce, it is a way to stay in close touch with his people and to know what's going on.

Through communication you can create an open and cooperative environment in your firm. Letting your employees know your goals for the company and giving them feedback on a regular basis creates excitement, promotes teamwork, and allows your people to make your goals their goals.

2.12   Being there

Your employees, subcontractors, suppliers, bankers, and customers see you as the "rock" of your firm in which they place their trust. Seeing you on jobsites and in your office, hearing you on the phone, knowing you're available when they need you—all these help keep their security index steady. Remember that your work ethic, involvement, and dedication set the tone for everyone around you. Habitually come in late, leave early, spend too much time out of pocket, and you will pay the price. At its best, your firm is a fine-tuned piano. Any neglect by you to maintain that is sure to spoil the concert.

2.13   Identifying objectives

You didn't start your business in a vacuum, having no idea what you were going to build or for whom. It's up to you to establish the identity of your company and determine the objectives you want to achieve. Be specific. A goal "to find new customers and make a good profit this year" tells neither you nor anyone else what they must do, and is not measurable. Instead, you've probably said something like, "This year, ABC Construction will increase total contracts by 15 percent through new customers, increase the gross profit margin by two percentage points, and limit general and administrative expense increase to one percentage point. To accomplish these, we will. . . ."

Once you've established your objectives, decide how you're going to accomplish them, and be sure your plan is up to the task.

2.14   Measuring results

Once you've established objectives for your firm, you can weigh them against real performance. You need to know how your actual general and administrative (G&A) expenses compare with your proforma budget each month (G&A expense is discussed in Chapter 9, "Accounting and record keeping") and you also need to know how your overall financial results for a given year fit with your intermediate and long-range projections. With clear goals that you established earlier, and current detailed financial reports, you can evaluate your firm's performance on any parameter. Looking at the bigger picture, is this year's performance consistent with your three-year objectives for the firm? Five-year?

Only with meaningful analysis and comparison can you make the mid-course corrections that are often necessary. And this kind of evaluation applies not only to financial performance but to every leg your organization stands on, including customer satisfaction, new customer development, quality of construction, reliability, your company's image, and employee performance.

When considering the "size" of your business, think not only in terms of contract volume but also—and more importantly—profits. It's a matter of keeping your eye on the ball. Focusing on contract volume can distract you from the number that matters: profit.

2.15   Marketing

As small general contractors do not usually have a marketing or sales department as such, this function probably falls upon you as the owner of your firm. Sales and marketing are discussed in Chapter 3, "Sales, marketing and business development."

2.16   Little habits with big payoffs

Here are a few simple habits you can easily develop that are certain to have a positive impact on your effectiveness.

- Seek to do business with and employ only people of quality and competence.
- Use an effective system for making and storing information on the spot, checking your calendar, making appointments, recording your brilliant ideas, etc., and keep it with you.
- Maintain confidentiality as appropriate.   No matter what they tell you to the contrary, very few people can manage sensitive information prudently (i.e., keep a secret).

- Mind your thoughts. A. P. Sherman, writing in *Entrepreneur*, quotes Paula Adkins, a senior executive at General Dynamics Corp., as saying, "Small people talk about people, medium people talk about events, and big people talk about ideas."
- Keep your word. This depends on two factors: intent, and committing to only what you can produce.
- Take time to think. This refers to two modes of thought: Focus on a specific issue or issues; and on unstructured thought in which your mind is set free. It's hard to do this in the rush of the day. Try it in your car or in your office at home. Your business success will be roughly proportional to the time you spend with external stimuli turned off and your brain in gear. Everything important begins with a thought.
- Listen before speaking.
- Stay close to your people. Know what they do and think.
- Be frugal, especially in the early years. Order of spending priorities should be paying taxes, building cash reserves, and paying down debt. Forever live beneath your financial means.
- Be sure your employees know the circumstances under which you are to be notified immediately, such as potential legal issues, significant customer disputes, events that may affect profitability, and public relations concerns.
- Read every day. Completion of formal training is not the end of learning but the beginning of a never-ending habit of reading that you must adopt. As a focused small-business person, you will find it almost impossible to read an issue of *Entrepreneur* magazine and not be motivated to action by one or several of its quick-read margin blurbs that address a number of business challenges or the emergence of a new trend that you may be able to tap into. You cannot plow through Stephen R. Covey's *The Seven Habits of Highly Effective People* and not be inspired to a higher level in your business and your personal life. Biographies of accomplished people scope out their thoughts, and teach innovation, cause and effect, problem solving, and other lessons of failure and success. We stay abreast of our field through industry-related publications, and of our world by reading newspapers and other general media. Reading is the way we keep up with the events and processes that influence our businesses and our lives. It is one of the ways we learn. A Chinese proverb says, "Learning

is a treasure that accompanies its owner everywhere." Without an effective reading habit, we cannot reach our potential.

2.17   Getting involved

In addition to industry advocacy organizations mentioned earlier, you can join other more general business groups. The National Federation of Independent Business (NFIB) is one of the most influential voices for small business in Washington, D.C., and your local Chamber of Commerce, as part of the umbrella Chamber of Commerce, lobbies for business in general.

Worthwhile civic organizations include Toastmasters, Rotary, and many others that provide the opportunity for personal growth and community involvement. Such organizations are common in industrialized countries.

These are only a few of the many groups available to business owners. Whether an active or more passive member, you will reap great benefit from belonging.

This chapter generalizes your responsibilities as owner of your construction firm, whether you delegate them to others or do them yourself. Other chapters go into more detail.

Chapter 3

Sales, marketing and business development

Nothing happens until someone sells something.

– Unknown

If you're a smaller contractor, don't make the mistake of thinking marketing and business development apply only to the big guys. A marketing program is needed to make your company known to your potential customers and give them reasons to do business with you. And even if you intend to stay small, you have to stir the pot of potential customers.

Start your marketing effort by knowing what construction market you're going to pursue, and learn it inside out. Specialization allows you to more quickly become known for the work you do.

Build your marketing program around your strengths. Your conversations, correspondence, and marketing materials all should succinctly convey to your prospective customer that "this is the contractor who can give me exactly what I need." As a specialist, you know who your potential customer is and you can reach him without the cost of marketing widely—i.e., to owners who don't fit your pattern. If you're just getting started and have no history to show your prospect, take heart. At one time, every contractor you know had zero projects under his belt.

Following is a description of the marketing materials, actions, and ideas that have worked for me.

3.1   Marketing materials

- **Brochure**   Prepare a color brochure consisting of a single page or a double- or tri-fold that fits in a standard business envelope. The brochure tells who you are and briefly explains why a prospective customer should do business with you. This is intended to present only enough information about your company to generate interest. Save the other details for later. Include a couple of carefully selected photos of completed jobs. Avoid hype and empty slogans, mottos, and mission statements. "We give the best in customer service" is fluff—words that sound nice but don't convey any concrete images. It's much more effective to point out something you've already done.
- **Letter of introduction**   Create a standard letter draft for prospective customers on your word processor that you can tailor as needed for the specific company you're contacting. Your letter takes over where your more general brochure ends, so take care to avoid excessive repetition (although some redundancy may be desirable to stress certain points relevant to the specific needs of the project owner you're sending it to) and limit it to a single page. Use a standard black font, white or off-white stationery, business-letter format, and a laser printer. Any graphics and color should be professional and businesslike. A sample introduction letter is included in Appendix 1.
- **Web site**   A well-designed Web site is an effective way to introduce your firm to prospective customers, as well as job seekers, bankers, and other business connections. It can include extensive information about your firm, and the reader can select the parts he's interested in. If you're not qualified to create a top-notch Web site, have it done by a professional, and be sure it's designed to get priority ranking on the major search engines in response to your own key words and presents a businesslike appearance. There are online vendors who can help with this, including www.topsitelisting.net and www.submit-it.com.

You can put a "customer portal" on the Internet that will give your customers round-the-clock access to information specific to that customer's project. Each customer can be provided his unique password for access to his page.

- **Contractor qualification statement**   Most owners require a formal statement of your qualifications. A form usually accepted by all parties is available from the American Institute of Architects (AIA). The AGC has a similar form. These forms require details about your firm's history, finances, key personnel, and subcontractors, so for convenience you can save the input data on your word processor and keep it updated.

- **Business cards**   Present just the facts. Avoid odd-sized cards, fancy typeface, and cards with a blank line for a name. And why not give every employee his own business card with his name printed on it? They are inexpensive, signify professionalism, and show that you value the employee.

- **Jobsite sign**   Place a company sign on your jobsites if allowed. It takes only one job generated by such a sign to cover the cost of all the signs you will ever place on jobsites in your entire career. You can have your name, phone number and colors printed on Tyvec® or similar weatherproof fabric and mount it on plywood. The cost is reasonable if you have multiple copies printed in a run. Be sure the sign is framed, installed plumb and level, and presents a professional appearance. Careless installation may be seen as an indication of the quality of your work.

- **Logo**   A logo becomes associated with your name and colors, especially if you use jobsite signs. Web-based logo design centers such as www.logoworks.com farm out logo creation to a number of graphic artists who will then submit original ideas for your review. You pay for only the logo you select, if any, and the cost of the whole process fits even the smallest budget. Do without a logo until you can afford to have one created. Avoid logos that may be used by others.

- **Prepared employees**   In small firms, the owner of the construction company usually heads up the sales and marketing effort but every employee is a sales and marketing representative. Each should be armed with a stash of the firm's promotional materials and be prepared to tell the company story in a few words whenever the opportunity arises.

In preparing your marketing materials, remember that your prospective customer's focus is on his own needs—*what you can do for him*. If you specialize in the kind of work he needs, make that clear in your brochure and letter of introduction. Edit your materials to limit history and detail about your company to what the customer wants to know in an initial contact. It's okay to have a lot of white space on the brochure, even an entire panel. You may feel that you're providing him less information than you'd like to, but the opportunity for that will come later.

If possible, hire a professional to help create your Web site, business stationery, brochures, business cards, handouts, jobsite sign, and logo with a theme of colors and design. This is an expense that will pay off in the image you present.

The grammar, spelling, and punctuation in your marketing materials must be correct. A prospective customer who has nothing other than your Web site or introductory letter by which to judge your company may view a misspelled word as a lack of attention to detail and wonder whether that might carry over into your work performance.

Send a complete marketing package and introductory letter immediately after you make a new contact—while your conversation is still fresh on his mind. Use customer relationship software to log your contacts and their data, and file a copy of all correspondence.

3.2   Publicity

Mass media advertising makes no sense for contractors, but you often can get free publicity in the local media with press releases that you or a publicity writer prepares and submits to them. Not only is this sometimes published at no cost, marketing people say it is remembered longer and is more effective than conventional advertising, partly because an article in the newspaper has more credibility than a paid outright ad. Learn which publications your target customers are most likely to read, and try to get a favorable article about your firm placed in one of them from time to time.

3.3   Proposals and presentations

Busy owners see unsolicited proposals as junk mail, so zero in on a few promising prospects in your target market and do some research before sending out your marketing materials. For each one, get the answers to a few critical questions: Does the owner bid all of his work, or negotiate some of it? With whom? How many projects does this owner build per year? What is his area of operation? What is his financial situation? What is his track record with general contractors? Who is the decision maker for the projects in your area?

Being knowledgeable about his company tells the owner you're diligent enough to have done some homework, and your chances of getting his attention go up. Be prepared to make the most of what could be your only shot. Concentrate on *his* needs—he does not care that your grandfather founded your company with ten dollars in his pocket back in 1946. Follow up with an email to address any questions or concerns he mentioned. Remember that price, your expertise in his field, construction quality, schedule, reliability, and effective relationships are likely to be his top priorities, even if unspoken. And nothing changes in the case of a corporate owner. You will still be dealing with an individual like yourself.

3.4 Staying ahead of the pack

As mentioned earlier, invest time in reading. General newspapers and magazines, business-oriented journals, trade publications, and newsletters specific to your construction interests are essential resources for effective business management. As just one example of their value, a publication may print a story about a company of interest to you who is planning a construction program in your area.

Businesses often announce their plans to enter a certain market many months or even further in advance, providing you the opportunity to contact them and maybe get involved at an early stage—before your less-informed competitors know anything about them.

Public companies often publish their development plans in their annual reports to shareholders. You can use these to check out one of your customer's or prospective customer's plans well in advance.

3.5 Impressions

The way you and your employees dress, speak, answer the phone, and put your thoughts in writing are just a few of the things that form the impressions you make on others. Most people feel better about themselves and their employer when the employer requires certain high standards as a condition of employment and enforces them evenly.

Many times, your front-office receptionist is a prospective customer's initial contact with anyone in your firm, either by phone or in person. Your receptionist should be prepared to make a good first impression—which often sticks in the mind regardless of what follows. For example, screening incoming phone calls can be offensive to the caller if not done right.

Your front-office people who may come into contact with visitors should make appropriate appearance, know the correct spelling and pronunciation of the name of everyone in the firm, have at least basic knowledge of your company, and handle themselves in an appropriately personable manner. The appearance of the reception area and any part of the office that visitors may see should present your company in the way you want it to be perceived.

3.6 New customers vs old

I've seen contractors go all out to attract new customers and at the same time treat their current customers with indifference. That makes absolutely no sense. It is easy enough to imagine that it costs a business much more time and money to win a new customer than to retain an existing one.

I've spent months and even years courting prospective multi-unit customers who held the potential for a long-term relationship. Once they become my customer I am not about to take them for granted, and to a large degree because of this, I had many of my customers' repeat business for many years. A work backlog consisting partly of repeat business and a portion of negotiated work are hallmarks of long-term success in the light-commercial construction business.

Keep a list of targeted prospects on your desk and spend some time with it every day—calling a couple of them, writing notes, sending a clipping they may find useful from a newspaper or trade publication, giving them important updates on your organization, etc. Keep a log of your interaction. If you have a newsletter put your customers and new prospects on the mailing list. However, avoid being overly solicitous or bothersome to any customer or prospect.

3.7 Reaching out

An important avenue to new business is through civic organizations like chambers of commerce, Toastmasters clubs, and Rotary. As an active member of your community, you can become more widely known and your name may show up in newspapers and business publications now and then. Sponsoring and participating in charity events such as golf tournaments will repay you many times over in terms of goodwill, referrals, and even direct business, not to mention the benefit to others.

Many people develop business relationships on the golf course even if business is never mentioned. According to Thomas J. Stanley, author of *The Millionaire Mind*, there is a positive correlation between net worth and playing golf.

3.8 Data mining

Dun & Bradstreet is a huge financial reporting firm that collects data from companies in various industries internationally, pools it, and sells it in many formats and configurations to other companies who need it in the operation of their own business. This service is not prohibitively expensive and may be useful in your marketing program.

Another source for customer prospects is magazine publishers whose target readers coincide with the focus of your construction firm. Publishers often rent subscriber lists to others.

Chapter 4

Creating customer loyalty

Customer Satisfaction is Worthless, Customer Loyalty is Priceless.

– Jeffrey Gitomer, *Customer Satisfaction is Worthless, Customer Loyalty is Priceless*

A t the minimum, your customer expects service and performance as agreed, quick response when you're called on, fair treatment, and privacy of business information. But simply meeting those minimum requirements doesn't go far enough. You must create loyalty. *Loyalty comes from consistently giving your customer more than is required or expected of you.*

I can't remember entering into a construction contract with an owner when we were not on good, even friendly, terms, each certain that our hopes for the project would materialize, but it didn't work out that way every time. This raised a fundamental question that I worked on over the years: What can a contractor do to improve the chances that both owner and contractor will be pleased or happy—i.e., more than merely "satisfied"—with a given project's outcome?

The answer is especially important to the niche contractor, who has great potential to develop an ongoing relationship with a chain store owner whose business strategy requires more or less constant generation of new operating facilities.

In my company, we reduced this answer to four factors: Budget, quality, relationships, and schedule. We put BQRS in block letters on in-house forms and used these factors as a starting place when discussing how a job was going in a regular staff meeting or when I was meeting privately with a project manager.

Here is a brief discussion of each of these four principles.

4.1 Budget

If the proposal you submit to the project owner is within his budget and otherwise meets his criteria, you're off to a good start. It's usually after this point that budget problems can begin to wear down the good relationship.

Costly change orders, even when completely justified, may throw the owner outside his internally approved budget. In the case of corporate chain stores, the owner's construction representative is expected to complete the project within the budget, and big change orders may indicate that he wasn't thorough in putting the bid documents together—an impression he doesn't want to be conveyed to his employer. Or if you're working with a chain store franchisee and change orders cause his costs to exceed his construction loan and reserves, he may unexpectedly need to go into his own pocket for the additional funds and this does not sit well with franchisees. In either case, the owner will look for places to throw blame for the cost overrun and you're going to get some or all of it—deserved or not. And it's likely to carry over to the next time this owner is selecting a contractor.

I say this to point out the reason that it is so important during bid preparation for you to anticipate problems that may crop up during construction and bring them to the owner's attention before the bid due date. Going this extra mile increases the owner's confidence in you.

Such diligence cannot always protect you. We once had under construction a free-standing chain store project on which we ran into a subsurface soil problem that wasn't apparent during a pre-bid site visit or indicated on any of the bid or contract documents. We stopped work in the affected area and notified the owner's construction representative who visited the site and signed a unit-price change order for the required remedial work, which ended up increasing the owner's construction cost significantly.

When the extra work was completed our invoice for the change order went up the line to the owner representative's boss's desk, who called my project manager and me in for a talk. He said he had been unaware of the magnitude of the extra work and felt blindsided. Even though he acknowledged that we had followed the proper procedures, his construction representative, who was on the jobsite while the corrective work was being done, had not kept his boss informed as the work progressed. Our customer paid the change order but we were tainted by association with a project that brought negative attention from above to the people who selected us to build the job, and they didn't involve us in their next few projects.

An error in judgment or quantity takeoff or unit pricing can put your profit margin at risk even before the project begins. If you're fortunate enough to learn of a problem early in the job, you have some time to try to work it out by forcing savings in other areas. Of course, you should be looking for those savings in any case, but you should take extraordinary steps when the chips are down. For example, some of your subcontractors may be willing to help make up for your mistake, especially in the case of subcontractors who do a lot of work for you. Also, your job superintendent may be able to tighten up on general conditions (job overhead) costs—an area, by the way, that requires continuous scrutiny.

In stark contrast to niche projects, the broader spectrum of general construction projects is often awarded on the basis of initial dollar cost to the lowest of a sometimes long list of bidders. They go so cheaply—to the bidder with the biggest error in his bid, some contractors argue—that the "winning" contractor has no leeway in his budget to give the owner anything extra in terms of quality and service that might gain him some goodwill with the project owner and, in fact, he may have no incentive to do so anyway because he knows the next project is again going to be awarded to the low bidder out of eight or ten or twenty bidders—regardless how well the current job turns out for the project owner. The contractor in this position must be eagle-eyed for every opportunity to claim compensation for extras outside the work he bargained for. This is tough, cold, hard building contracting that often exists in an atmosphere of adversarial relationships between the contractor and the owner or his architect (and doesn't sound like something I want to do every day). Control of factors that affect budget are discussed in more detail in other chapters.

4.2 Quality

One of the important advantages of specializing in a certain segment of chain stores is the opportunity to achieve a *preferred* status with a busy company, thus more mutually supportive working relationships than those described in the above paragraph. Many chains are typically very conscious of the image their stores present to the public, which means they place high priority on the quality of their buildings' finishes. This expectation effectively eliminates from competition the contractors who are unwilling or unaccustomed to turning out front-room finishes.

This is not to say that any good general contractor cannot meet such expectations, but to do so some of them would need to develop new subcontractors, retrain or hire new employees for certain positions, and make adjustments in operations. There's also attitude: Many contractors and their employees stiffen their necks at an owner's demands for what the contractors see as finish quality that is a cut or two above construction industry standards. Be thankful for these contractors, as they make more room for those of us who accept such owner demands as part of the dues for membership in an exclusive club.

The standard for construction finishes required by many chains is excellence. And as motivational writer Ralph Marston said, "Excellence is not a skill. It is an attitude." True, if accompanied by a skilled craftsman.

Chain store contractors and their project managers are accustomed to some give and take with the project owner representative, both parties presumably representing their respective company's best interests. A spirit of cooperation usually exists. The project owner representative may be overly picky about certain things that he focuses on, but he may give a valued contractor some slack on others.

The term "quality" has different meanings to different parties to the construction project—the owner, contractor, architects and engineers, subcontractors, and others. Aside from the quality of structural integrity as defined in the contract documents, which must not be compromised, the meaning of "quality" varies from industry to industry, project to project, contractor to contractor, and from one owner to another.

The requirement for excellence extends well beyond the jobsite. An owner's perception of the level of quality a contractor produces is certain to be tarnished by that contractor's sloppy handling of paperwork, poor communication, and incomplete documentation. The project owner has performance obligations as well but, of course, you can control only that of your own organization.

In order to achieve a completed project that meets the owner's quality expectations, all parties to a project must acquire an understanding of those expectations, incorporate them into the contract price and other contract documents to the extent possible, and commit in good faith to carry them out. You as the contractor have a responsibility to yourself to be certain this part of the overall arrangement is given the required attention in advance of signing the construction agreement.

4.3 Relationships

People do business with people they like. It's human nature. Face it—you do this in your personal life as well as your business relations. If all other factors are equal, likeability will control. Sometimes it controls even if other factors point elsewhere. There can be any number of reasons one person may take a liking to another, but it's safe to assume that the elements of character and personality would make most lists. Most people also respond to others who show sincere interest in their personal well-being as well as their business success. Writer Dale Carnegie said that our individual success comes from helping others achieve their goals.

This is never more true than in your relationship with the project owner's representative who is your direct contact. If your performance makes him look good to his superiors, he is likely to find a way to keep you on his "A" list of contractors. If your work makes the project owner look good to his customers or his tenants or his stockholders, you will be the first contractor he thinks of for his next project.

With repeat business, the contractor's benefits include the learning curve and a more predictable workload. The owner benefits from a contractor who has proven himself and requires little orientation and hand holding. By working as a team on the project, each party can reduce legal expense and time involved in negotiations over contracts and minimize the likelihood for disputes. And each party's comfort level increases with every completed project.

Your relationship with the project owner depends heavily on your reliability. During both the construction and the warranty period, the owner wants to make one and only one call to you if something you are responsible for needs attention, and he doesn't want to feel he must follow up to see if it has been handled. An unattended problem reflects on his judgment in choosing you as his contractor, a mistake he will not want to repeat. In general, the owner wants the following with regard to reliability: *Do what you say you will do when you say you will do it.* And know that a given problem does not just "go away." You will fix it now or you will face it again later, along with an unhappy project owner. The positive feelings generated among all the parties by a smooth construction project can be wiped out after substantial completion if the contractor's response to final punch-out, clean-up, or warranty issues is absent, slow, or carried out with a poor attitude.

Never lose sight of the fact that construction is a service business. The project owner's primary objective is for his construction project to stay beneath his radar problem-screen so that he can tend to his other responsibilities. As much as we contractors may want to believe otherwise, construction to the project owner is not an end in itself. It is merely a vehicle he uses to accomplish his overall plan—leasing the building to others, occupying if for his own business use, or profitably reselling it. Your job as his contractor is to provide a level of service that enables him to meet his goals.

Occasionally a project owner or his representative comes along that is not worth the hassle involved in doing his work. He may ride you too long before making payment or make totally unreasonable demands that may affect employee and subcontractor morale, and job profitability. Once a new construction representative for a major chain I had a long-term relationship with called me to his office to discuss his rules for his contractors. One of them was a no-change-order rule. When I asked for clarification, he said there were simply no conditions that would warrant extra payment. I declined to work for him. It is an absurdity to say there will be no change orders regardless of circumstances. Such a policy sets the stage for an impasse when something unforeseeable comes along, such as unstable soil below the surface. Things returned to normal a short time later when the company sent in a new construction representative.

4.4   Schedule

Today's chain operators are acutely aware of the time value of money and demand optimum construction duration in order to get a return on their investment at the earliest date. The keys to optimizing schedules are an organized project owner whose staff and custom materials suppliers can meet tight schedules; an experienced, skilled, streamlined contractor with motivated project managers and superintendents; cooperative and capable materials and equipment vendors; and experienced subcontractors who are willing and able to perform tightly-scheduled work.

Tight schedules face obstacles. Stretched-out regulatory inspection departments often require several days advance notice. If you try to anticipate when you will be ready for the inspection and schedule it too early, you go to the bottom of the list if you're not ready when the inspector shows up. Project owners are more often providing many of the building components. Take Starbucks for example: They provide even toilet fixtures and other common items—leaving the contractor with less control over scheduling. In tight labor markets, subcontractors may be unable to properly staff your project. And developing nations intermittently strain the world supply of building materials.

Many contractors use positive follow-up systems as part of their quality assurance program. Some send a questionnaire to the project owner or arrange a meeting with him shortly after completion of a project and again after six or twelve months for this specific purpose. Your objective is to gain a picture of your organization from the other side, which you can use to improve your service to this customer. Conducted properly—without unnecessary accusation or defensiveness by the parties—such follow-up tells the customer that you are attuned to his concerns and likely teaches your staff how to better serve this and other customers.

As described here, Budget, Quality, Relationships, and Schedule (BQRS) represent the fundamentals that will enable you to build loyalty in your customers. Just keep raising the bar in all four areas.

Chapter 5

Business considerations

The man who complains about the way the ball bounces is likely the one who dropped it.

– Lou Holtz

5.1 The corporation

For purposes of this writing, it is assumed that you will form a corporation as your business vehicle. A corporation comes into existence upon issuance by the state of a corporate charter, which describes its business scope as submitted by its organizers. The scope should be broadly stated to include activities that the corporation may want or need to undertake in the future.

As a legal entity a corporation stands alone, independent from its owners. It is authorized to do business, enter into contracts, buy, sell, and own real estate and other types of property, and borrow money.

A corporation may sue or be sued in its own name. Corporation owners (stockholders) are generally protected from the debts and liabilities of the corporations they own, including those resulting from a lawsuit, but as owner of a closely-held corporation, i.e., one whose shares are not traded on an open market, you can lose this insulation, one of the primary reasons for forming a corporation, if you fail to maintain strict separation between yourself and the corporate entity. You might do this, for example, by using the corporate bank accounts for personal expenses. Learn from your CPA or tax attorney how to maintain this separation and set up procedures to protect it. If you fail to do so, your creditors and other claimants may be able to look beyond the corporation to you personally to satisfy their claims.

The two most common types of corporations are known as "S" corporations and "C" corporations. An important difference between the two is that S corporations do not pay federal income tax directly. Instead, an S corporation's profits flow through the corporation to the stockholder(s) and are reported on the stockholder's individual income tax returns. Certain restrictions exist for S corporations, but they offer distinct advantages for some small building contractors.

Stockholders and directors of both S and C corporations should hold annual meetings and keep minutes of these meetings with their corporate records as a legal formality. More information is available at www.irs.gov. Talk with your business lawyer and your CPA to learn the relative advantages of each business structure as they might apply to your company.

Limited Liability Companies (LLCs) are similar to S corporations with regard to taxes and personal liability protection, but are not a good choice for a construction firm.

5.2   Capital equipment

Contractors have learned the hard lessons of owning expensive capital equipment. Chain store builders and other light-commercial contractors usually subcontract out practically all of the work to others, who own or rent the larger machines and tools they use. Few checks are harder to ink than a monthly payment for a backhoe that is sitting idle, as it is sure to do from time to time. Instead of buying company vehicles, contractors often pay superintendents and project managers for the use of their own pickup trucks for company business.

Maintenance of capital equipment is a problem. It's hard to assign responsibility to one person for the proper care and maintenance of a machine that's used by everyone unless you have a full-time maintenance employee, which for a small firm isn't practical. Repair issues creep in, and one day when your loader or dump truck is urgently needed on a jobsite it's down. At other times it's needed on two or more jobs at once.

Capital equipment depreciates rapidly. Maintenance and repair are costly. Transportation to and from jobsites is a logistical problem and expensive.

Insurance rates are high. Accounting is more onerous when you own equipment. A secured yard or building is required for storage.

These are just a few of the reasons to resist any urge to purchase capital equipment.

5.3 Purchasing

How you purchase, receive, and care for construction materials directly affects job profitability. Effective procedures and forms are needed to buy at the best prices, receive orders as specified, and check delivery tickets against orders. Upon receipt of the materials, a responsible employee of your company must check the delivery to be sure it matches the original quantities and specifications, and effectively deal with any discrepancies as necessary to maintain the schedule and preserve your rights for adjustment by the vendor or shipper.

Delivery receipts should be examined on front and back for any language that transfers undue responsibility to the contractor. A Georgia contractor was held to a provision on the back of a concrete delivery ticket, which the contractor's job superintendent had signed, that the contractor would indemnify the supplier for all damages for personal injury or property damage that was not the result of the supplier's *sole* negligence. When one of the supplier's employees was killed on the jobsite, the contractor was held liable for the supplier's negotiated settlement with the killed employee's family. Proper education and training of your employees are required to reduce the possibility that you become saddled with liability for which you are not responsible.

Whenever possible, suitable terms should be negotiated with each vendor and included in a written agreement or purchase order that takes precedence over field delivery tickets and other documents. Jobsite employees do not always have the time or expertise to recognize such transfers of risk to you.

In general, title for a purchased item remains with the seller until delivered to the purchaser if the purchase order or other agreement requires the seller to deliver the item to the buyer's location. However, there are exceptions to this rule.

5.4 Collection

Project-owner payment to construction contractors is frequently slow. This may be due to inadvertent delay or intentional foot-dragging by the owner to improve his cash flow or reduce his interest expense on a construction loan.

Whatever the reason, slow payment is a source of much contractor angst and is a significant contributor to cash-flow problems or worse.

Slow payment by the project owner may not be eliminated but you can take steps that are sure to help. At or before signing a construction agreement, meet or talk with the project owner at the highest level possible in his organization. In the case of an individual owner or small franchise operator, meet with the owner himself. If a larger company, learn who has final check-approval authority and meet with him. Also include in this meeting the project owner's representative you or your project manager will be dealing with throughout the course of the project. A good time for this meeting is before or during execution of the contract by the owner and the contractor. In addition to signing the construction documents, a primary purpose of this meeting is for you to politely but firmly, eye-to-eye, stress your expectation of payment as specified in the agreement, just as the owner expects you to perform on the agreed schedule. Especially on short-term projects, a commitment sealed with direct eye contact and a firm handshake is not easily forgotten. Such personal commitment can be more effective than the written terms in a construction contract. The very first time a payment from the owner is late, address it with him—reminding him of the earlier commitment he made to you. Knowing you will quickly call his hand will likely cause him to pay you before he pays others whom he knows are less "squeaky." Be diligent and consistent about this. However, some chain owners, even financially healthy ones, hold onto contractor payments for sixty days or even longer. Reasons for this include financial problems, sloppy handling of paperwork, and improvement of cash flow. If slow pay is simply the policy within a larger company, your best efforts may not get you any special treatment, in which case you have to decide whether your cash-flow circumstances can handle it, and whether you're willing to put up with it in order to get the business. Of course if a project owner develops financial problems, you may need to discuss a course of action with your construction attorney, and possibly act quickly. Try to get to know the owner's payables clerk—an advantage when you call to ask the status of a payment you're expecting.

When the project owner is using a bank to finance a project, as part of ensuring that you will be paid you should speak or meet with the bank officer handling the loan. In most cases the bank will understand that you and the bank have a mutual interest in all parties to the project getting paid.

Ask the banker for a letter stating that the bank will reserve loan funds sufficient to pay you the amount of the owner–contractor contract plus a reasonable amount for unknown extra costs. Otherwise the pot of gold from which you are to be paid can be depleted because the owner may use the funds for other purposes before you are entitled to receive final payment. An owner who has to dig into his personal pocket for the money to pay you can be much tougher in negotiating change orders and making timely final payment than if he were using loan funds. It's a good idea to keep in touch with the bank officer throughout the project.

In the owner's process of completing a construction loan, his bank may ask you to sign an agreement to complete the project if the owner defaults on his loan. If you agree to do so, you should require a provision in that agreement that the bank will pay you up to date for the work you have completed at the time the bank takes over—*before* you are required to begin work for the bank.

At times it is necessary to place a mechanic's lien on the owner's property to improve your chances of collection from a non-paying owner. This course of action has legal implications and requires an attorney familiar with lien statutes. Be aware of time limitations and other state law requirements for placing liens before starting a job in that state. Lien laws vary from state to state.

Liens are discussed in more depth in Chapter 8, "Building It."

5.5 Dealing with the IRS

As a businessman for more than three decades, my only out-of-the-ordinary dealings with the Internal Revenue Service (IRS) have been limited to one "audit" and probably a dozen letters from the agency requesting additional information about an item on my personal or corporate tax returns.

The audit lasted less than an hour. It was held at the office of my CPA, who provided the requested records. The IRS agent was satisfied after examining several of them and made no changes in my tax return. As to the letter inquiries, I cannot remember even one that my bookkeeper or CPA did not resolve with a phone call or by providing a clarifying document to the IRS.

Here are a couple of factors I believe can at least reduce your chances of an unpleasant encounter with the IRS:

- Convey to your bookkeeper and your outside CPA your intent to comply with tax laws and to meet filing and payment deadlines without exception. This in no way precludes you from scouring the tax codes for every legal means of reducing your tax liability.
- Use a CPA who remains current on the ever-changing tax laws and court rulings, has no important blemishes on his reputation, and is recommended by his clients. Professionalism and reputation are attributes that are not likely to be ignored even by an IRS agent, who after all is a human being and who has some latitude in his handling of your situation.

Nevertheless, it is easy to draw unwanted attention from the IRS. For example, there are limitations on how much salary you as the owner of your firm can draw. A contractor I know was investigated by the IRS for taking what the agency declared excessive compensation over a period of years. The IRS's claim was for millions of dollars for unpaid taxes, interest, and penalties. The contractor won the dispute after a two-year legal battle but spent more than $200,000 defending himself in addition to his cost in terms of distraction from running his business and mental anguish.

There's one IRS factor that is beyond your control. The IRS annually swoops down on a small number of supposedly randomly selected taxpayers and examines every morsel of their tax returns for a certain year. Even if these unfortunate taxpayers' income tax returns are 100 percent clean, their businesses and/or personal lives are disrupted. This kind of IRS audit is called the Market Segment Specialization Program (MSSP). Your chances of being one of these unlucky taxpayers are extremely low, but the program is something you should know about.

The IRS provides for a free wealth of information at its Web site www.irs.gov. From there you can go to areas of interest such as construction.

5.6 Contractor failure

It has been said that construction is the riskiest business you can get into. If so, what are the reasons?

Anyone who has met all the regulatory requirements can call himself a contractor and attempt any project an owner is willing to give him. The new contractor often is someone who has achieved some success as project manager, superintendent, or executive in another construction firm, and has decided to do it on his own before he has acquired sufficient capital and management experience. Being a good project manager does not directly translate into being successful at running the business of construction.

Some states license general contractors, but a license does little to ensure good business management. Licensing authorities usually impose certain financial standards, but a contractor's income statement and balance sheet are a snapshot taken at an earlier date and may have changed for the worse. There are some advantages to licensing, and some states' licensing requirements are more effective than others', but licensing is no guarantee of a contractor's viability.

Licensed or not, a hungry contractor may be tempted to take on a project that's outside his realm of competency and financial capability. An astute project owner (through his architect, CPA, or attorney) can do a much better job than any licensing authority is likely to do in determining a contractor's qualifications to build his project, including length of time in business, financial strength, qualifications of key personnel, experience with projects like the owner is awarding, favorable recommendations from the contractor's earlier clients, current workload, and the contractor's experience in the locale of the project under consideration. Softness in one of these areas may or may not be significant, but should be considered in light of the project at hand because it can result in problems for both contractor and owner, ranging from minor to catastrophic. But unfortunately, for both owners and contractors, contractor background checks are often superficial.

A contractor who starts with too little capitalization, hoping he can generate some profit before he runs out of money, is out on a limb: For instance, the owner can be slow in paying, or the job could turn out to be unprofitable; meanwhile, the contractor's expenses may eat up the limited cash he started with.

The state of the economy is often given as a reason for contractor failure. "Bad times" is not a very good excuse because economic downturns are a certainty and good management practice means having a plan to deal with such predictable events even if their timing is unknown. But if the economy tanks early in the life of a contractor who hasn't had to manage through hard times in the past or who hasn't built up cash reserves, he is vulnerable. Many contractors fall into this category and some fail. The contractor's greatest risk is during his first few years in business.

Always be aware of your vulnerabilities and be prepared to take the action necessary to save yourself from a non-recoverable situation. For example, you desperately want to keep all of the employees who are so valuable to you, but when times are bad you have to weigh that against your need to survive the storm intact and rebuild when it's over, which means making some very difficult decisions about payroll and other expenses.

Be aware that the job costs and overhead expenses will keep coming through the pipe long after the valve is cut off at the source and can eat up the reserves you're counting on to see you through the problem you're faced with. Weigh this lag factor when timing any downsizing.

No one has better described the reasons for contractor failure and how to avoid them than Thomas C. Schleifer in his book *Construction Contractors' Survival Guide*. The opening sentence in his preface summarizes his philosophy:

> **Your management decisions alone determine whether you will succeed or**
>
> **fail in the construction business.**

Schleifer knows. As a consultant to the surety (bonding) industry for a decade, his work with contractors in trouble afforded him an internal view of the circumstances that brought on their problems and the management decisions behind them. Prior to being a consultant, he was a general contractor.

I call attention to Schleifer because he had an important positive impact on my business. In his book, he identifies ten major areas of risk and I have used it often in employee training. His lessons undoubtedly steered me away from more than one risky decision.

Schleifer does not say the identified areas of risk are to be avoided—only that they must be approached with awareness and proper planning. That advice has a broader application as well: Don't allow the risk to keep you out of the construction business, just approach it sensibly and responsibly.

Chapter 6

Controlling your finances

Experience is a hard teacher because she gives the test first, the lesson afterward.

– Vernon Law

Although you may (and must) employ someone qualified to keep your books and a CPA to guide you and your bookkeeper in establishing proper accounting procedures and tax planning, you yourself must remain in close touch with your finances.

6.1 Working capital

The United States Small Business Administration (SBA) says lack of adequate working capital is one of the top causes of small business failure. In addition to having enough cash for day-to-day operations, you need reserves that will enable you to survive the tough times. In a large corporation such as GE, a design flaw that results in the mass recall and modification of thousands of one of its major products may cost the company many millions of dollars, but that isn't likely to threaten the viability of such a large and diversified company. But in a small construction firm, especially a young one, making a substantial bidding error or going an extended period of time without securing profitable work can have consequences ranging from merely worrisome to catastrophic. So build up cash reserves as early as possible, and establish a line of credit to provide funds for temporary needs (but don't use loans to support a losing operation).

Establish relationships with more than one bank. You will probably use one more than another, but if you ever need the second bank, you will be ahead by having established some history there. Banking relationships and loans are discussed in Chapter 13, "Banking and Finance."

6.2   Projecting cash needs

Theoretically, you should be able to operate a construction business with little cash. After all, you're going to pay your subcontractors and other vendors on a thirty-day basis, aren't you? And your customers promise to pay you within thirty days, do they not? So all you need is enough cash for miscellaneous expenses and payroll and your construction draw will come just in time to pay the subcontractors and Home Depot, right?

What's wrong with that picture? Even with diligent cash management, we all know that things don't work out quite so neatly. How much cash you need on hand and how much is available from loans depend on many factors. A cash forecast is an attempt to project your cash needs at a time in the future by estimating your cash inflow and outflow. On the outflow side are G&A expenses including payroll, rent, insurance, and taxes; and job costs including payments to subcontractors and suppliers. Cash inflow usually amounts to owner payments for the construction draws you submit.

Timing payroll, rent, and other G&A outflow are relatively easy, but subcontractor and vendor payments are as random as job starts, and receipts from owners are often delayed for any number of reasons. This amounts to a complex cash flow picture that requires updating at least as often as the schedules for your new projects are known.

Taking on more work than can be supported by your working capital, management team, and other infrastructure is as likely to cause failure as being unable to acquire work. Other financial responsibilities that fall on you include establishing budgets for a chart of accounts that you've tailored to your business. You must monitor accounts receivable, accounts payable, job costs, overhead expense, cash balances, bank lines of credit, payroll tax remittances, income tax payments, and other important gauges, and act quickly when one of them gets out of line. Establish reporting formats that your bookkeeper will use for each of these indicators, and review these reports on a regular schedule. (Some of these are included in the weekly "Vital Signs" report described in Chapter 9, "Accounting and Record Keeping.")

6.3 Understanding financial statements

Unless you already have a good understanding of how to read and interpret income statements and balance sheets, ask your CPA to spend some time teaching you at least the fundamentals. Consider enrolling in related seminars and classes. This is important for comprehending your own financial reports and those provided to you by project owners, subcontractors, and other vendors. Financial statements are crucial tools in your decision-making process. But when using them to evaluate others, be aware that anyone can throw dreamed-up numbers onto an un-audited financial statement, so all statements should be viewed with a critical eye (yours or your CPA's) when their accuracy may affect your well-being. High-profile accounting scandals in giant corporations have shown that even audited financial statements must be viewed with caution.

6.4 Dishonest employees

An employee who is determined to cheat will find a way. Even an "honest" employee who happens upon a weakness in the system may be tempted and eventually steal from you. Your CPA can help you devise systems that at least make cheating difficult and more likely to be discovered.

A project manager or field superintendent may on occasion be offered a favor or outright payment from a subcontractor or vendor in exchange for something in return from your company that your employee controls. Your employees should understand that accepting bribes is a firing offense. Be careful about making accusations without the proper proof and consider consulting with an employment attorney about such a matter.

A fraud sometimes used by dishonest employees is through the establishment of fictitious vendors or subcontractors, a scheme found in construction and other industries as well. Let's say you have a project manager ("Jack") who is authorized to approve vendor invoices submitted to your company. Jack covertly sets up a fake company, XYZ Plumbing, which then issues invoices for non-existent charges to one of your actual projects that Jack manages. When the phony XYZ invoices arrive at your office, Jack approves them and they are processed for payment the same way as for the invoices of your real vendors. Finally, your firm's check arrives at the XYZ post office box owned by Jack.

Another situation to warn your employees against is a request by the project owner for you to invoice entities other than the owner named on the construction agreement or to otherwise manage payments and invoices in an unusual manner. Doing so could be innocent or it could have serious consequences for your firm. Any such transactions should be made openly, in writing, with your knowledge, and only with the advice of your own attorney if you are not certain as to the potential consequences.

The above are only a few of the many avenues for dishonesty or mishandling in construction. You can never be guaranteed that it will not occur in your company, but a good "firewall" is diligence in hiring, effective training of your employees, proper controls, and your careful oversight of vendor payment.

6.5   Where is the money?

Be sure you understand your internal flow of cash receipts and disbursements. In a small business you may know all of your vendors personally, but this becomes less likely as you grow larger. One simple control to put in place is for you personally to look for your signature on all cancelled checks. You can usually do this online or on a CD provided by your bank.

Chapter 7

Bidding

It is a reality of life that men are competitive and the most competitive games draw the most competitive men. That's why they are there—to compete. To know the rules and objectives when they get in the game. The object is to win fairly, squarely, by the rules—but to win.

– Vince Lombardi

A thorough treatment of bidding would fill an entire book and is not included here. The purpose of this chapter is to highlight a few areas of risk and opportunity for your consideration when putting a bid proposal together.

Restaurant and some other chains do not have formal public bid openings. Instead they may require delivery of all bids to their office on or before a specified date and time, after which they open them without the bidders present. The owner is not required to, and often does not, make all of the bids known to the bidders. Chains generally feel free to award a project to the bidding contractor of their choosing based on their own internal criteria.

Chains whose business model typically includes generating a steady stream of facilities in which to do business limit bidders to contractors who fit their mold and use them repetitively in the interest of reducing hand-holding time and receiving consistent quality in their projects. Because of contractor attrition and other factors, contractor selection is an ongoing process for chain operators. Specific contractor selection is generally the domain of the owner's construction representative—and rightly so, because he will be the person held responsible for the contractor's performance. You may make inroads with an owner or his representative who is interested in developing a new contractor, but he may not award you the first job you bid for him. However, he may give you some meaningful feedback about your proposal and encourage you to keep trying.

7.1   Qualifying to bid

Each state requires entities doing business in that state to follow certain prescribed registrations prior to conducting business activities. This includes registration with the secretary of state and possibly other agencies, depending on the state. Simply bidding on a project may constitute activity that puts you subject to the secretary of state's registration requirements or the rules of a contractor licensing board, so you must anticipate when you plan to bid in a state and investigate its requirements in time to comply. Counties and cities may have additional requirements, including obtaining a business license.

Many states require bidders and contractors to also obtain a professional license prior to bidding or building a project within the state. The licensing process can be arduous and require weeks or months of time, and the penalties for non-compliance severe—including not being allowed to use the courts to secure payment for your work.

It is also important to research well in advance the lien laws in states you plan to work in. Lien laws vary from state to state and some require certain notices by the contractor prior to commencing work. Failure to strictly comply with applicable lien statutes can disqualify you from placing a lien, or may render your lien unenforceable.

Registration and licensing and mechanic's liens are discussed in more detail in Chapter 8, "Building It."

7.2   Approach to bidding

If you're a building contractor you're probably an optimist, but realism is required when putting together the prices for projects you bid on. You may have a reliable price for every building trade and all materials, but not everything in the course of building the project will go as planned. This becomes apparent when, say, your concrete subcontractor stops performing in the middle of the job for some reason and delays the schedule, and you pay a premium to get another subcontractor to take over the concrete work. On the other hand, if you throw too much money into the bid for the unexpected, it is very likely to be spent.

Let me talk about this last point briefly and then I will return to bidding. You should require accountability for each line item of cost in a project's budget, not just total project cost against total budget. This eliminates any tendency your project manager may have to be satisfied with letting line-item savings balance out line-item cost overruns. Gains are necessary to improve your profit margin, which you may have reduced for a competitive bid situation. Losses must be accounted for. Only if a line-item loss is identified as such can its cause be determined and steps taken to avoid it in the future. Otherwise it is perpetuated and over time becomes factored into future cost estimates and job budgets. This cycle affects your competitiveness and ultimately your viability as a contractor.

Long-surviving contractors are line-item aggressive, managing projects with the expectation of improving on the bid price for each line item. Earning the budgeted profit on a project should never be considered a satisfactory outcome.

Now, back to bidding. A common cost underestimation is in job duration. On short-term projects, the time required to mobilize, complete the final punch list, and demobilize may easily amount to 10 percent of the construction time, but this may be overlooked in bidding.

Another cost variable that's sometimes hard to predict is downtime due to bad weather, holidays, etc. The direct overhead job cost to maintain a jobsite on even small projects is many hundreds of dollars per day and your profit on the job is reduced by that amount for every lost day. Anticipate a reasonable number of unproductive days based on local conditions and circumstances, and include them in your estimate of dollar cost and contract time. Then be sure that schedule is met or beaten.

7.3 Pricing

As a chain contractor you're likely to include subcontractors on your bid list who are not only accustomed to working with your people, but also have experience building the brand you're bidding on. The combination of your experience and theirs with a certain chain gives you a high degree of confidence that the subcontract prices you use in your proposal will be thorough and competitive, the "unknown" factor in bidding a first-time project having been overcome.

If you're not getting at least one out of four of the projects you bid on competitively, take a look at your bidding practices, materials cost, subcontractor prices, and general conditions cost. The defensive posture of blaming a poor win rate on other contractors who're taking jobs "too cheap" stands in the way of your own improvement. There's an old construction tale about a contractor who's just learned that he lost a bid by exactly one dollar: As he stomps away from the bid opening, he whines that the lower contractor cannot possibly build the job for that price.

If you're well entrenched in a niche market, your hit rate should be better than one out of three or four competitive bids. Add to that the higher rate of negotiated jobs associated with specialization, and your cost of getting jobs should go down and your work load and profit margins up.

7.4   Cost databank

A cost-coding system used diligently by your field employees, project managers, and bookkeeping staff can track your cost down to the cost of nails used in framing if you were to want that much detail. Each store you build of the same type provides you the opportunity to further refine your database and consequently your ability to submit as tight a bid as you think is necessary to get the next similar job. Another advantage of accurately job-costing identical buildings is that a comparison may help you spot any significant cost deviations in the course of a project while there's still time to make adjustments.

7.5   **Pre-bid site inspection**

The owner should provide a topographical survey of the site that shows the lot corners, existing buildings, site surface elevations, finish floor elevations, manhole elevations, storm and sanitary pipes, utility poles, fire hydrants, trees, and other objects on or affecting the site. It should show the elevation relationship between fixed points on the site and adjoining properties.

Where the topo survey is intended to show existing features, the site plan shows what is proposed. Together they are expected to give you and your subcontractors and suppliers all the site information needed to prepare an accurate site bid.

Surveys, site plans, soils reports, and other bid documents can contain errors or omit important information. In such cases—and this is where contractor mistakes are easily made—if the bidder does not catch an error that a thorough inspection of the site would have revealed, any cost consequences of the discrepancy that are discovered after the construction agreement is signed and the project is underway may fall on the contractor.

If no soils report is included in the bid documents for a project that includes site work, include a disclaimer in your bid to the effect that you are not responsible for any soil conditions not suitable for the proposed site improvements, foundation design, and the like.

Here are some of the factors to consider in the course of a pre-bid site inspection:

- Completeness of the bid documents. You may be responsible for obvious site conditions that are omitted or inaccurately shown on the bid documents.
- Access to the site. Traffic patterns or other factors may dictate certain times of the day that materials deliveries and subcontractor access are not feasible. Logistics becomes a factor in projects that are constructed inside other buildings—e.g., building out a space inside an office building or hotel; more so if not on the street-level floor of the building. Consider any time involved in moving all of your equipment and construction materials to and from the jobsite by way of an elevator, especially one that will also be used by the building occupants during construction.
- Surface drainage. Soil conditions and provisions for drainage of surface and runoff water from adjacent properties affect drying time. Non-porous soils may require several days after rain to become dry enough to permit site work activities.
- Weather norms. In some areas and seasons, weather conditions can be a significant factor in projecting cost and construction time.

- Security. Some urban locations in high-crime areas require fencing and manned security throughout the construction period, sometimes around the clock.
- Use of adjoining and nearby properties. Judge whether the business and other activities on nearby properties could affect when and how you carry out your work.
- Adjacent structures that may be affected by the work required by the proposed project.
- Evidence of rock, seeping water, unsuitable soil, surface drainage from adjacent property.
- Availability/proximity of subcontractors, material suppliers, equipment rental sources.
- Environmental issues.
- Parking space for employees, subcontractors, material suppliers, and others.
- Availability of the utilities that will be required during the construction process.
- Relocation of utilities. Relocation of a heavily loaded power pole can be time-consuming and very expensive. Know who pays the cost. Obtain relocation cost figures from the utility owner. Don't guess.
- Labor union requirements if any.
- Placement of storage/field office facilities.
- Other factors as dictated by the project at hand.

The above should not be considered a comprehensive list for any given project.

You can minimize the risk of error or oversight in bidding the site work by developing a broad-based checklist to be used on all pre-bid site visits. Make the site inspection early enough in the bid process to allow time for running down any differences you discover between the bid documents and the actual conditions. And never make the assumption that a site is so simple and straight-forward that no site visit is necessary. Site work holds tremendous opportunity for both profit and risk.

Estimating site work cost is a ripe area for competitive advantage. As no two sites are exactly the same, cost data is more or less incomparable from one project to the next. This fact provides you the opportunity for creativity in bidding the site work, more so than may be possible in pricing the building itself when the building is a prototype that the other bidding contractors are familiar with.

On sites that require moving dirt to or from the site, there can be huge differences in the bids you receive from your grading bidders. This is due to several factors, including

- A grading contractor's workload at the time of bid
- The grading contractor's success in locating suitable fill material or a dump site nearby, since it is usually the haul distance and not the cost of the material that most affects his cost and his price to you

Grading contractors who are willing to do small jobs are not always the most professional. The more site work bids you receive, the more likely you are to get one from a contractor whose circumstances mesh with your needs. Be prepared to go to your second choice when the job starts if a low bidder is not available.

Require your site work bidders to submit their proposals to you in the form of an itemized price list for the various elements of the site work, as you've defined them, so that you can check any given bid for gaps and overlaps and so that you can directly compare all of the site work bidders' prices against each other. Otherwise, small grading contractors may bid in a format that suits themselves instead of according to bid instructions.

7.6 Warranty considerations

Consider your potential liability under the warranty provisions in any bid documents. The warranty may be limited only by the applicable statute of limitations and could run indefinitely. Your potential liability must be covered in your bid if the owner will not allow reasonable limitation of the warranty period.

7.7 Compiling your bid proposal

Preparing a bid for submission to a project owner or architect requires much careful work in addition to determining your bid price. Here are only a few of the details you need to observe.

Chain stores often provide bidders a proprietary standard form on which you are required to submit your proposal (sometimes electronically). To minimize the chance of error, establish your own standard forms and procedures for compiling bids and follow them consistently. When you're required to use the owner's bid form, reallocate the numbers from your system to the categories on the owner's form and cross-check the forms' totals.

When you receive bid documents, check the actual drawing and revision numbers against those on the title page or the list provided by the owner or architect. When you submit your bid, list the drawing numbers and revision dates that your bid is based on, including any revisions you received after the initial package. Finally, if you're awarded the job compare your list of bid documents to those included in the owner–contractor construction agreement. These procedures prevent disputes or costly errors that might occur from a mix-up in versions of drawings.

Some chain store bid procedures permit you to include comments in your proposal. If so, you should clarify uncertainties or omissions in the bid documents, exclude certain work such as correction of soil conditions, and note special instructions you've received from the owner that may not be in the bid documents. If comments and clarifications are not allowed in the proposal, discuss any such issues with the owner prior to the bid date and confirm any agreement in writing. Don't rely on oral agreements for this; put them in writing. The person you've agreed with may be transferred or no longer work for the project owner when a related question arises. Also, memories can be short.

Occasionally one of your customers may ask you to work up a price for a miscellaneous job for which no drawings exist. In this case, write a detailed "Scope of Work" along with sketches or drawings as necessary as a basis for pricing and construction, and get the owner to approve it. This scope of work should be incorporated into your construction agreement with the owner.

Before submitting proposals to owners, screen them against your own standard final checklist.

7.8 Reverse bidding/auction

A few private owners are experimenting with the reverse auction concept by which all bid prices for a given project are posted on a designated Web site, followed by a second and sometimes further rounds of bidding. When the bidding period ends, the job is awarded to the lowest final bidder.

Unless it becomes standard industry practice, I personally would not participate in this kind of bid, which puts the procurement of construction services on the same level as the purchase of a commodity such as ready-mix concrete, which can be completely defined by standard specifications. I don't believe reverse auction will survive to any significant degree in the United States. In fact, the US Army Corps of Engineers has decided against the use of auctions for its projects.

Chapter 8

Building it

Everybody has a will to win. What's far more important is having the will to *prepare* to win.

— Bob Knight, *Knight: My Story*

Overall field operations are beyond the scope of this book but it would not be complete without a look at some of the overlaps between field operations and business management. (An excellent book on construction field operations, also written by a contractor, is *Construction Operations Manual of Policies and Procedures*, by Andrew M. Civitello, Jr.)

8.1 Registration and licensing

Many states require professional licensing of general contractors, and the direct and consequential penalties for non-compliance can be severe. Where a professional license is required, the applicant usually must take a written examination covering state law and administrative regulations as well as general and state-specific construction knowledge (e.g., earthquake, hurricane), and meet the state's financial criteria. It may take several weeks or months to obtain a professional license, and, as mentioned earlier, some licensing authorities require bidders to have a license even *before submitting a bid*. If you violate these laws and regulations, you may very well run into trouble. Among other consequences, you may not be allowed to use the state's courts to sue for non-payment. This does happen. Operating under another contractor's state license may subject you to even worse problems. The risk is not worth it.

Even in jurisdictions that don't require professional licensing, you'll have to register your business with the secretary of state or another governmental arm.

This tells the state you're working there and enters you on its tax rolls. You also must obtain a local business license.

The AGC has simplified the problem of determining the requirements for each state in the United States by publishing its *State Law Matrix*, which provides state-by-state information about the statutes that relate to the construction process. The Matrix is available by subscription and only in electronic form through the AGC. It includes information about the various states' pre-qualification requirements, bidding requirements, lien laws, payment clauses, subcontracting requirements, and other information, and is updated periodically. For more information or to purchase this service, go to www.agc.org and search for "State Law Matrix."

In addition to states, many cities and counties have their own codes and licensing regulations. Don't guess at the regulatory costs and tax liability in your own or another state you plan to do business in. Investigate them before signing an agreement to perform work in any state, as they can be significant.

8.2   Environmental studies

A prudent property owner or project owner will have conducted environmental studies on the project site, but you should satisfy yourself that this has been done. Environmental laws have long tentacles, and just about anyone having worked on a property that was found later to be contaminated may be held responsible even though the contamination occurred prior to his presence on the property. Environmental insurance protection is limited. Be sure to discuss this with your insurance agent.

8.3   Subcontracting the work

Fewer and fewer general contractors perform work with in-house forces, instead subcontracting most of the work to subcontractors who furnish their own labor, common materials, tools, and supervision. Outsourcing eliminates the need for layers of supervision at your jobsite and frees the job superintendent to manage the project through subcontractors' supervisors. It permits a leaner contractor organization all the way up the line, reducing the direct administration and overhead costs associated with the larger numbers of contractor employees that would be needed to work on jobsites.

8.4 Photographs

Not only is a picture worth a thousand words, it may save you lots of dollars. Shoot a set of pictures of the jobsite after you're awarded a contract but just prior to beginning work. Stand at corners of the building site and snap pictures along the property lines. Shoot the structures near the property line on adjacent properties to show any damage that exists and the condition of sidewalks, curbs, trees, etc. Also snap enough photos to show all apparent features of the work site itself including any structures.

Take photographs throughout the course of construction showing general progress, site conditions after any weather events that impact your schedule, and other significant conditions. Photos should be part of your documentation of any changed or differing conditions claims, and change-order work as well. Use digital cameras that imprint the date and time and archive the photos. You will refer to them more often than you think, and in the event of a dispute or damage claim you'll be glad you have them.

8.5 Pre-construction meetings

The benefits of conducting a pre-construction meeting that brings the major subcontractors and suppliers together include the opportunity to clarify construction details, review the schedule, and discuss working arrangements among the various players. Your subcontractors and other vendors become acquainted and learn from each other's questions and comments. The job superintendent may lay out the jobsite rules for all to hear and discuss at once, stress commitment to the schedule, and go over safety rules and accident-reporting procedures.

Commitments regarding schedules and cooperation made at these meetings by all relevant parties in good faith, face-to-face, are less likely to fade in the course of a short-term project. If a subcontractor or supplier fails to perform as agreed after making such a commitment, you are in a better position to jawbone him back into compliance than if you were relying solely on the contractual provisions. Persuasion is likely to result in a better outcome than the blunt force of a written agreement, although you should always have one. I usually worried that I had a serious problem if I had to start reciting contract language to a subcontractor.

Insist on attendance at pre-construction meetings even in the face of the many reasons you will hear from one party or another that he cannot attend.

8.6   Before you start a project

There's a lot of pressure to get a project started once the owner and contractor reach agreement in principle, but first be sure all the i's are dotted and t's crossed. Requirements vary from job to job, but here are some of the contractor must-haves, based on my experience:

- All necessary permits and entitlements (including right-of-way, development, land disturbance, environmental, building, etc.)
- Satisfaction that environmental studies were conducted and show no indication of contamination (usually arranged by the project owner)
- Registration with the local secretary of state
- Professional contractor's license where required
- Local business license
- Construction agreement executed by owner and contractor
- Formal Notice to Proceed from the owner if required by the owner–contractor agreement
- Confirmation that the owner's financing for payment to the contractor is in place, if applicable
- Insurance coverage including builder's risk, commercial general liability, worker's compensation, and other insurance required by the construction agreement and as otherwise prudent
- Verification that the owner has in place all insurance he's required to provide, that your company is named as Additional Named Insured on his policies, and that all parties have waived subrogation rights (see Chapter 14, "Insurance and Bonds")

- Copy of the policies of all other parties' insurance coverage that affects you
- You should be shown as Additional Named Insured on all subcontractor policies
- Subcontracts signed and delivered
- Insurance certificates from all subcontractors *before* they begin field work on the project
- Location and flagging of underground utilities by the local Call Before You Dig center

Serious problems that can be prevented by thorough preparation include laying the building out incorrectly on the site or placing it partially or entirely on the wrong site. Avoid this by requiring the owner to stake and identify all corners on the site. Do not take this responsibility upon yourself. As a precaution, some contractors hire an engineer or surveyor to verify the site corners placed by the owner. You may also choose to have an engineer lay out the building corners.

See a more complete pre-start-up checklist in Appendix 1.

8.7   Project overhead/general conditions expense

Project overhead costs are the costs required to carry out the work of a project but that cannot be assigned to any one item of the work. These costs are determined on an item-by-item, job-by-job basis but may be condensed into a single line item in your bid and budget. Alternatively, some contractors use their historical project overhead costs to determine a flat percentage that they use on all cost estimates, but this is not ideal due to the variability in overhead requirements from one job to another.

The requirements for a specific project vary but the typical small project includes the following overhead items:

- Barricades
- Builder's risk insurance and other special insurance
- Building materials unloading and handling
- Cleanup
- Demobilization

- Drinking water
- Dumpster
- Equipment unloading, handling, assembly
- Erosion control
- Field office supplies
- First aid
- Job superintendent salary and expenses
- Jobsite office and storage facilities
- Jobsite signage
- Legal expense related to the project
- Mobilization
- Municipal bonds (guarantees to government agencies that you will comply with their requirements)
- Photos
- Project manager (prorated among all of his concurrent projects)
- Protection of existing structures and landscaping including those on adjoining properties, easements, and right-of-way
- Provisions for special weather conditions
- Security fencing
- Security services
- Temporary driveway/parking area
- Temporary utilities, including electrical power, toilet, phones, water
- Testing
- Tool and equipment rental
- Travel expenses including temporary lodging
- Worker's compensation insurance (unless included elsewhere)

The costs associated with many of these items are a function of time. On short-term projects, delays caused by unusual weather, late equipment delivery, etc. can have a relatively larger impact on job time and project overhead than on longer-term projects since there is less time for plusses and minuses to average out. While you may be entitled to a time extension for such a delay, you're not likely to be paid for its associated cost. You may be able to make a successful claim for delay costs if the delay is caused by the owner.

8.7.1 Managing project overhead/general conditions cost

Among the project overhead items listed above, those that require and best respond to effective project management are

- **Payroll** If a job goes beyond schedule, you lose not only the extended payrolls but also the opportunity for your superintendent, and to a lesser degree your project manager, to produce profits on other projects.
- **Tool and equipment rental** A Bobcat that is needed for only a few hours can sit around accruing rental fees for days after it's no longer needed unless the job superintendent remembers to call the rental agency for pickup. A few hundred bucks here and there resulting from delayed returns quickly add up to thousands in the course of even a short-term project. You may be creative enough to devise a plan that will reward the superintendent for tight management of rental costs (e.g., he is paid a percentage of any unspent portion of the general conditions budget).
- **Purchases of small tools and supplies** Set up a system that forces separation of expendable supply cost such as fuel for space heaters from the cost of durable tools that includes saws, drills, portable heaters, etc. An employee purchasing a durable item at company expense should be required to account for it, and your bookkeeper must track and inventory it.
- **Cleanup** Control your cost for site cleanup by imposing and enforcing a contractual requirement that your subcontractors clean up after themselves, or share in your cost.
- **Travel** The geographic boundaries of a traveling contractor's operating area are often beyond the distance the job superintendent can practically commute to and from his home on a daily basis. Establish daily values (per diems) that you pay the superintendent for lodging and meals on out-of-town jobs, and include their total cost in your bids and cost budgets. Also, establish as policy how often you will pay for the superintendent's cost of travel between his home and the jobsite (e.g., every weekend, alternate weekends, monthly, etc.). You may establish different per diems for project managers, whose travel patterns are usually different from a superintendent's.
- **Project schedule** When you look at the above items that go into job overhead it's easy to understand that their combined cost can approach a thousand dollars per day for even smaller light-commercial construction projects on which the contractor's only direct employee is the job superintendent. Good

project management dictates critical path scheduling that identifies the earliest possible date each task can be started. If followed diligently, critical path scheduling leads to optimum construction time. If not, a day or two or more of delay creeps in here and there and affects the overall schedule. As a job that is allowed to proceed without critical path scheduling nears the completion date, the project manager and/or superintendent begins to feel the heat and tries to make up the lost time by overloading the job with trades that must work on top of each other even more than usual on such projects, resulting in cost inefficiencies, frustrated subcontractors, reduced work quality, and all-around anxiety by all parties having a stake in the completion date. Some of the lost time may be made up this way, but at a steep price in terms of subcontractor and employee dissatisfaction and frustration. On repetitive projects, job-familiar superintendents and project managers may forgo computerized scheduling yet still manage to meet tight schedules. As a successful contractor, you must manage to consistently meet your internally projected schedules, which usually should be earlier than the contract completion date. The biggest factor in doing so is commencing each project "task" at the earliest possible date and completing it on schedule.

Be prepared for owners to challenge you on inclusion of project overhead in your bids, pointing to the overhead portion of your Overhead & Profit (O&P) line item. You should be able to overcome their objections with a thorough explanation of the content and reason for each type of overhead. In their simplest terms, overhead in the O&P line item is the cost of running your business operation; project overhead/general conditions are the cost of running the job.

8.8   Warranties

Consider your potential liability under the warranty provisions in any bid documents. Negotiate for favorable warranty terms, including a time limit on your warranty obligations. Otherwise your warranty may be limited only by the applicable statute of limitations and could run indefinitely.

Once under contract, begin to address the warranty requirements in subcontracts and project management procedures. Require subcontractors and suppliers to submit warranties and guarantees in accordance with the contract documents as a condition for payment.

Equipment manufacturers may be able to deny warranty coverage if their equipment is not installed in strict accordance with the manufacturer's specifications.

In addition to express warranties written in the contract documents, there are also implied warranties. For example, there is an implied duty by all parties to a construction project to deal fairly with each other and to act in good faith. Another example of implied warranty is an owner's warranty of the adequacy of the drawings and specifications he provides to the contractor.

8.9   Mechanic's liens

A lien is a security interest in property. For construction purposes, it is a legal means available to material suppliers, workers, subcontractors, and general contractors to encumber an owner's real estate when the owner does not pay as agreed. Lien laws vary widely from state to state, but a common element is that the terms of the lien statutes must be followed precisely if a lien is to be valid. (To be safe, you almost certainly need a lawyer who regularly uses the lien statutes to place a lien.) You should research the lien laws of any state in which you build a project, and structure your payment procedures and documentation accordingly so that you will know how to protect yourself from liens placed on the owner's property by others, for which you may be responsible, and what you must do to protect your right to lien the owner's property if that should become necessary.

In some states, the maximum amount for which a subcontractor may place a lien is the unpaid balance of the owner–contractor agreement. This protects the property owner (who may or may not be the project owner) from double payment; in other states, the owner of the property on which the project is built may be liable to lienors even if he has already paid the full contract amount to the contractor.

Some states' lien laws require the material supplier or contractor to give formal advance notice to the owner that he is going to perform work on his property. Failure to do so may jeopardize his right to lien the property. In still other states a lien in favor of the contractor or supplier is automatically placed on the property upon commencement of the work. When placing a lien, be aware that your lien is usually secondary to any mortgages and other claims placed at an earlier date. A lien must be filed within the time limit specified in the lien statutes of the various states. Beware of contract provisions that attempt to limit your right to file a lien.

A lien secures your interest for a period of time, but further legal action is required to collect the amount owed—possibly an expensive proposition that you must weigh against the range of possible outcomes. Collection also depends on other factors, including your ability to prove your claim in court and the value remaining in the property after claims having higher priority than yours are settled.

Bankruptcy by the owner changes everything. In this bleak event, check with your construction attorney before placing a lien, terminating your agreement, or taking other actions against the project owner to collect payment. Not only are your actions invalid in some instances, they may subject you to bankruptcy court penalties and attorney's fees.

The lack of uniformity in the lien statutes of the various states makes it impossible to put them into a neat, easy-to-use package. The above general considerations are meant to alert you to the significance of liens and prompt you to consult with your construction lawyer about the lien laws of any state you work in and the steps you need to take to protect your interests.

8.10   Lien waivers

A waiver of lien rights (a "lien waiver") is usually required by contract from the general contractor and all subcontractors and sub-vendors as a condition of owner payment. (Your contract with the owner should provide that these waivers are not required for the payment currently applied for, which can be a particularly difficult requirement, but only for prior amounts the owner has paid you.) To protect yourself from erroneous or false claims that you did not pay your subcontractors and vendors, set up procedures to collect proper lien waivers from each of them and from lower tiers. Know your direct subcontractors' and vendors' suppliers. Since even a single missing waiver may hold up your application for payment, or leave you open to claims, your diligent administration of lien waivers and other payment documentation is required.

Make each payment to your subcontractors and vendors conditional on your receipt of properly executed lien waivers for the payment applied for or the prior payment, as required by the subcontract. State this condition in purchase orders and subcontracts, prescribe the forms to be used, which may vary from state to state, and require the subcontractor or vendor to deliver the properly signed waiver to your office in exchange for his check.

8.11   Closing out the project

When a project is physically complete, the project manager and superintendent may be anxious to move on to the next project, but get the current one completely behind you first. I think it is better to keep the superintendent on the job until all punch lists are complete, but it is just as common for a contractor to send in a different crew to punch out.

Put checks in place to ensure all closeout items are done timely so that they don't become orphans. Final payment from the owner to you is conditioned on completion of certain closeout items, just as your final payment to subcontractors and other vendors must be. Here is a partial list of closeout responsibilities applicable to many types of projects. A convenient closeout checklist is included in Appendix 1.

- Administration of outstanding claims
- All punch lists completed and signed off. The final punch list should be designated as "Final" and signed by the parties when it is created and again upon its completion. Any legitimate issue that comes up after that time falls under warranty and should not be allowed to hold up final payment
- All systems tested and operated
- Certificate of occupancy in hand
- Final accounting and reports to management
- Final billing to owner
- Final claims waivers/releases from vendors and sub-vendors
- Final cleanup

- Final lien waivers/releases from vendors and sub-vendors
- Final payments to vendors
- Jobsite records delivered to the office including logs, test reports, field change orders, inspection placards, as-built drawings, lien waivers and releases, punch lists, subcontractor back charges
- Office, storage facilities, contractor signs, mailbox removed
- Project insurance discontinued as appropriate
- Rented items returned
- Retained documents organized by destroy dates and archived
- Set up mail forwarding to corporate office
- Temporary utilities discontinued
- Warranties and manufacturer's owner manuals and documentation to owner

Chapter 9

Accounting and record keeping

Accounting problems arise most often not from the accounting systems, but from the information fed into them.

– Thomas C. Schleifer, *Construction Contractor's Survival Guide*

There is no function more critical to your success in construction than effective management of financial and other records. Only with good accounting and record-keeping policies and procedures can you achieve the following functions:

- Collect the information you need to monitor the performance of current and completed projects
- Manage payables and receivables
- Forecast cash needs
- Improve your estimating function based on historical job cost
- Tag the best-performing and non-performing employees
- Identify unprofitable customers
- Expose dishonesty
- Comply with tax laws, while taking every advantage permitted by them
- Provide information required by various government agencies
- Know the financial status of your company at any given point in time
- Archive records that will be required for various purposes in the future

9.1   Certified public accountant

Professional expertise is required to design the record-keeping systems. Hire an outside CPA for this purpose. Once the systems are in place they can be maintained by a competent in-house bookkeeper or bookkeeping staff and reviewed periodically by your CPA. Your CPA is a consultant to you and your book-keeper, who may also be a CPA.

The complex and changing nature of tax laws demands CPA oversight and input. However, construction accounting is unique and you should seek out a CPA who has extensive experience in the construction field. Investigate his professional expertise and reputation and ask to speak with some of his clients. Consider his personality. Even the smartest CPA in town is not automatically a good fit for your company.

9.2   Audited financial statements

A set of audited financial statements bears the written and signed opinion of a certified public accountant, whose reputation and credentials are thereby at risk. The standards CPAs follow must be high so that lenders and other users of audited statements can confidently rely on them in making financial decisions related to the entity the audit represents. If a bank, for example, makes a loan to a company on the basis of its financial statements that turn out to be inaccurate, the bank may sue the CPA or CPA firm who put its signature on those financial statements. If you must have audited statements as a requirement of your bonding company or bank, be prepared for your CPA and his staff to be probing. Although it's you who's paying them, they have a professional duty not only to you but to themselves and to anyone who relies on their opinion of your financial circumstances. However, when not in conflict with their professional duties to others, you should expect them to place your interests first.

Who requires audited financials? Your banker may require them. Your bonding company certainly will. Some owners may.

9.3   Bookkeeper

Your bookkeeper is a key person in your firm. While knowledge and experience in construction bookkeeping are strong pluses, he needs a variety of organizational and management skills that range beyond strictly bookkeeping. What are these skills?

- Ability to learn new things quickly
- Ability to see an overall picture as well as the details
- Ability to organize himself and his work
- Interpersonal skills for dealing with his own staff, company management, subcontractors, vendors, banks, etc.
- Communication skills
- Quest for excellence
- Effectiveness under pressure

9.4 Cash vs accrual accounting procedures

Your CPA will recommend the accrual method of accounting. Under this method, income is taken into account in the fiscal year in which it is earned, which may be different than that in the period in which it was received. Similarly, expense is recognized when it is incurred whether or not paid during the period.

9.5 Percentage of completion vs completed contract reporting

9.5.1 Percentage of completion method

This method recognizes income on a project as the job progresses. This is done by comparing the actual costs incurred with the estimated total contract costs (cost-to-cost). Good construction software computes this automatically. This means that a portion of the profit on jobs that are not completed at year end will be reported in that year. This is the generally accepted method for reporting project income for financial statement purposes as well as for income taxes. For example, if one of your projects is 80 percent complete at the end of the fiscal year based on expected costs, then 80 percent of the projected profit is reported in that year.

9.5.2 Completed contract method

For income tax reporting there is a "Small Construction Contracts Exception" in the IRS code, which allows the profit on each job to be reported in the year the contract is substantially completed. This means that if you qualify for this exception you can defer the profit on jobs that are open at year end. The obvious benefit is deferment of income tax on the open projects. A Small Construction Contract is generally one that is expected to be completed within a two-year period and is performed by a contractor whose average annual gross receipts for the three years preceding the year in which the contract is entered into do not exceed a certain annual limit ($10 million as of the date of this writing). It is perfectly legitimate to report income on uncompleted jobs in your financial statements while deferring it for income taxes if you qualify for this exception.

9.6   General and administrative expense

In contrast with job-specific or project overhead discussed earlier, general overhead is the cost necessary to keep the doors of your business open, which cannot easily be allocated to any one project. General overhead is commonly called general and administrative expense, or G&A, or simply overhead. G&A includes office rent, heat and air conditioning, electricity, phones/broadband, computer equipment, software, office supplies, furniture, executive and administrative salaries and expenses, outside accounting fees, legal expenses, subscriptions, advertising, non-job insurance, and similar items associated with operating your business—i.e., just about every expense that would continue even if you had no jobs under construction.

As an example, let's assume you project your construction firm's current year G&A to be $490,000 on a business volume of $10,000,000, or 4.9 per cent projected G&A. So, if your total bid price for a certain job is $900,000 and includes your overhead and profit markup of, say, $105,000, it's realistic to assign $44,100 of your annual G&A to that job. This forces you to recognize that the resulting net profit before taxes on the job is $60,900—not the $105,000 number shown on the Overhead & Profit line of your proposal to the project owner.

9.7   Fixed vs controllable G&A expense

A construction firm ordinarily has two kinds of G&A expenses: controllable and fixed. Many controllable expenses may fluctuate greatly from one accounting period to another and often can be modified or even stopped in relatively short time. This is the first place to look when circumstances force you to cut expenses quickly, and also when you find your general overhead creeping up.

Controllable or variable G&A includes

- Salaries. Base pay and overtime for officers, administrative, and other employees not charged to a project
- Payroll expenses. These include unemployment taxes, social security taxes, paid vacations, worker's compensation insurance, sick leave, and health and life insurance paid by the company
- Advertising and marketing
- Vehicle and travel expenses
- Accounting and legal
- Repairs and maintenance
- Office supplies
- Outside services (includes temporary labor)

Fixed G&A expenses are those G&A costs that remain relatively constant and cannot be readily modified from month to month. These include

- Rent
- Utilities
- Insurance
- Loan interest
- Depreciation

9.8 Cost accounting

Cost accounting is the systematic assignment of the cost of time, materials, and overhead associated with the various elements of job cost to a set of codes for accounting purposes. The use of cost codes is meant to enable management to monitor and review costs associated with any part of a project, e.g., the cost of the reinforcing steel in a floor slab.

Cost coding is used to follow the progress of a project and to compare each line item in a cost budget to the final actual cost. Field cost codes should duplicate the codes on the cost estimate.

Continuous review of job cost reports is a critical project management function throughout the course of a construction project. A problem line item can be flagged while there may be time to make corrections.

Historical cost coding is valuable for bidding, especially in the case of repetitive projects such as are common in the construction of chain stores that use the same or similar design for multiple locations. It also serves as a benchmark against which the outcome of future projects may be measured.

The success of a firm's cost coding system is dependent on the following:

- The design of the coding system itself
- The balance between coding too much and too little detail
- The diligence and accuracy of each employee responsible for assigning codes to job costs
- Capable accounting software
- Timeliness of data input
- Analysis and effective use of the resulting reports

Cost accounting is not a separate set of books, but merely a categorization of data.

9.9   Financial statements

A company's set of "financial statements" are simply the various pertinent reports derived from the company's books of account. Together, these financial statements, if truly representative, present a complete picture of the company's finances from which an accurate opinion of the company's overall performance and financial position may be determined.

A formal set of financial statements shows a company's profit or loss and other financial barometers during a month, quarter, or the firm's fiscal year, and its resulting financial position as of the last day of that period. You will use your financial statements to review and make adjustments in response to past performance, and to project future workload capacity, borrowing and bonding needs, and staffing levels. They're required by outsiders like your bank, bonding company (always), materials suppliers, insurance companies, other vendors, and the owners you do business with.

The components of your financial statements that you're likely to become most familiar with are your income statement and your balance sheet.

9.9.1 The income statement

The corporate income statement shows the profit or loss during a given reporting period—month, quarter, year, etc.—as the difference between the income and the expense recognized during that period for all the projects combined and including G&A expense. Be aware that the income statement is a picture of the reporting period, which is history by the time you read the report, and does not show the status of your current operations; nor should it be used as an indication of the company's future performance. An income statement covering the most recent six-month period may look rosy overall, but may not reveal a downward trend that began during the last month of that period and is leading off the new period in a loss situation. Reporting on a rolling thirteen-month basis as discussed later minimizes this uncertainty.

9.9.2 The balance sheet

The balance sheet is a summary of the assets and liabilities of a business at a given point in time, usually the end of a month, quarter, or the company's fiscal year. For privately held corporations, as most construction firms are, financial statements are the property of the firm's owner, who usually provides copies only to the company's associates who require them in the course of doing business with the company, such as its lenders, insurance carriers, bonding company, and licensing authorities.

A company's assets consist of anything of value it has purchased, including jobsite equipment, tools, cars and trucks, office furniture and equipment, real estate, cash, accounts receivable, securities, and prepaid expenses.

Liabilities include unpaid loan balances, accounts payable, leasehold obligations, accrued expenses and taxes, and provision for income taxes.

The net worth of a firm is its assets minus its liabilities. Assets and liabilities are shown as "current" (to be realized within one year following the balance sheet date) and "long term" (the portion of an asset or liability that falls beyond one year).

Ratios of various elements of the income statement and the balance sheet are an indication of a company's performance and should be seen by the owner as pointers to steps he may take to improve his company's viability. Your CPA can help you establish ideal ratios as points of reference.

There are many good resources for understanding financial statements. Keep in mind that construction accounting is unique. As a business owner, you must acquire the ability to understand your own financial statements and those of others whose financial status affects you.

9.10   Reports

The construction accounting software you choose will probably be capable of generating more reports than you will ever want. Your CPA will require the reports he needs for review and tax preparation, and identify others that will be required by you and by your bookkeeper.

Here's what I like the reports to show currently about each of the company's projects separately:

- Initial contract amount and projected cost
- Change order price to the project owner and estimated change order cost on a current basis
- Final adjusted contract amount and total cost on completed jobs, for direct comparison to original projections
- Status of billings on current projects
- Aged receivables
- Amount of payables and due dates
- Original and projected or actual completion date

The above is a basic "contract status" report which I like to see on a weekly basis. If your software does not generate an at-a-glance report that includes all of the above, your bookkeeper should be able to quickly compile it for you via spreadsheet. Make sure the data from your jobs is current.

In addition to contract status, devise a one-page, weekly "Vital Signs Report." This can be in the form of a simple spreadsheet with columns for Cash on Hand, Accounts Receivable, and Accounts Payable as of the current Friday or Monday. Another column, which I call Vital Signs Index, is calculated from the above and shows the net sum of your cash and receivables less your current payables. Each week the current values for each column are listed on the next row of the spreadsheet. This report reflects changes in key indicators from week to week, reveals trends over longer periods of time, and is useful in anticipating cash needs and cash placement. The report contains nothing new, just a distillation of data from other sources that your bookkeeper can quickly compile, but it provides a lot of information in its simplest form. A couple of minutes with your Vital Signs Report on a Friday afternoon may assure you a peaceful weekend, or may hint at an approaching need for cash while you still have time to deal with it. This report will help prevent your being blindsided by an unexpected cash-flow crisis.

The Vital Signs Report can be expanded to provide other useful information. You might want to know your billings amount for the week or the total amounts of incoming and outgoing checks. But include only information you will use. Anything you add takes more of your bookkeeper's time. A sample Vital Signs Report is shown in Appendix 1.

Another important indicator is total G&A expense. Unchecked, G&A creeps up quietly and dangerously, rising along with your workload but not by a constant ratio. Avoid a surprise at year end by poring over your G&A line items monthly. They are certain to fluctuate from month to month and you must satisfy yourself as to the cause. If any expense item cannot be justified, slap it down before it becomes entrenched. Once a year, look at each G&A line item on an absolute rather than comparative basis.

Looking at your business on a rolling last-thirteen-month basis rather than the standard year-to-year and year-to-date basis allows you to not only compare your firm's current monthly performance against the same month last year, but also gives you a monthly look at your most recent period of any duration you wish.

9.11   Billings

Construction contracts generally provide for periodic payments by project owner to contractor. In commercial construction, payment at monthly intervals based on the estimated percent of completion is common. The construction agreement should specify the conditions the contractor must meet in order to be paid.

At the beginning of a project the contractor submits to the project owner (and/or architect, if one is involved) a schedule of values for each line item in the project cost breakdown. The sum of values of all line items plus the contractor's markup is the lump-sum contract amount.

The exact percent of completion of a partially complete line item, e.g., plumbing, can be hard to determine and a careful project owner will scrutinize the payment application to try to assure that the contractor or a subcontractor has not inflated the value of completed work. The owner wants the unpaid balance of the contract amount at any time to be sufficient for completion by another contractor if the original contractor fails to complete the project. (In reality, this is hard to achieve due to the difficulty in accurately determining the percentage completion of the work and because any replacement contractor would justifiably price such a project cautiously due to the difficulties involved in taking it over mid-stream.)

You have your own reasons for not inflating the percentage of project completion. If you bill the owner for more than the completed value of a line item and pass the payment down to the subcontractor but the subcontractor never completes the work, you the contractor will have to absorb the cost of having it completed by others.

9.12   State sales tax

If the state in which your business is based has a sales tax, you will pay the tax on all taxable materials purchased within that state at the time of purchase (unless you have a tax exemption, in which case you pay the accumulated tax later). But if you buy materials from outside your state and import them, you are required to file a Sales and Use Tax return and remit the tax. Your home-state tax collector, upon periodic audits of your books, will bill you for the tax amount you would have paid had you purchased the materials in your home state, offset by any tax you paid to the state of purchase.

The same procedure takes place when you build a job in a state other than yours. The foreign state's tax auditors will periodically come to your office and determine what materials you and your subcontractors imported into that state, and require you to pay its tax rate on untaxed or under-taxed imported materials, offset by any tax payments you've made to that state. You should know the foreign states' sales tax procedures.

States perform sales tax audits at differing intervals. Be sure your bookkeeper has a system for keeping up with the amount of sales tax you owe to each state you work in, recognize it on your books currently, and pay it when due. Rest assured that the tax collector will come and will be thorough in auditing your books. Sales tax collectors do not like to leave empty-handed. If you fail to recognize the sales tax you owe on a timely basis, the accumulated amount can be a jolt to your bank account and Profit and Loss (P&L) statement when it is due.

Chapter 10

Contract terms and conditions

Be prepared.

– Scout motto

At the time of execution of a construction agreement, the owner and contractor are typically friendly but that atmosphere can go downhill in the course of a dispute, and unless satisfactory resolution is reached it may evaporate completely. This highlights the need for a comprehensive contract. This chapter discusses only selected contractual issues that in my experience demand special attention when it comes to signing a construction agreement and is not intended to provide a comprehensive list.

Unless you are skilled in interpreting legal documents you'll need to hire an attorney who specializes in construction law to review any new-to-you construction contract and go over it with you before you sign it. The contents of this book are not intended to provide legal advice.

Terms that are important to you may be absent from a contract proposed by the owner, or it can be loaded with provisions that may have serious consequences for you—whether intended or not. Before executing any construction agreement, understand your rights, duties, and the risk it places on you.

Obviously, laws applicable to the construction process vary from state to state and from country to country. The laws of the United States and/or its political subdivisions shape the contract provisions referred to here. However, many of them are based on practicality and may be useful to contractors across state and national boundaries.

10.1   Types of agreement

There is no universally standard owner/contractor agreement form used in the United States. The most widely used family of construction agreements are those published by the American Institute of Architects (AIA). The Associated General Contractors of America (AGC) also publishes forms that are commonly used. Contractors in the United Kingdom, Europe, Canada and other countries should inquire at their respective construction organizations about commonly used contract forms.

The AIA and AGC documents offer the advantage of well-thought-out standard provisions that are periodically fine-tuned on the basis of industry experience, changes in the law, and for other reasons.

There are important differences in the AIA and AGC documents. Here are a few of them:

- The AGC documents place the architect in the background. This allows for more direct resolution of routine matters between the project owner and the contractor, which is generally favored by contractors.
- AGC documents make the owner responsible for interpretation of the contract documents. The AIA documents give this authority to the architect but do not necessarily hold him responsible for his decisions, a situation contractors don't like.
- The AGC documents remove the architect from disputes between owner and contractor.
- The AGC documents integrate the general conditions into the construction document. The AIA general conditions are in a separate document.

Since no form document is likely to address all of the concerns of all parties to a contract, some modification is usually required, and the terms and conditions specific to a given project must always be added. But the use of the AIA and the AGC sets of documents eliminates the need for creation of every word and provision of a proposed document. The included provisions may be accepted by the parties, or they may be changed or deleted as agreed.

Chain operators often require the use of proprietary contract forms prepared by their own attorneys. These almost always favor the owner who created them and it should concern you if the owner refuses to make reasonable changes. Once you've hammered out suitable contract terms with an owner, you should be able to get them included in future contracts with that owner routinely. Owners may make changes in their standard forms from time to time, therefore you should check each new contract for revisions before signing it.

Never assume a form that appears to be an AGC or an AIA document is genuine or unaltered unless you're certain of its source. Parties have been known to make document revisions in their favor without informing the other party and this can be done with visual perfection.

10.2   Requirements for a binding agreement

Lawyers caution that for a construction contract to be legally enforceable the following conditions must be present:

- Agreement between parties as to the meaning of the contract at the time of execution
- The parties must have the legal capacity to form a contract
- Sufficient consideration, or value

Unwritten contracts created by oral agreement may be enforceable but the complexities of construction, the probability that the parties may fail to include all of the necessary terms and conditions in their oral agreement, imperfect human memory, and even the possibility of "convenient" memory argue against the use of oral contracts. Unwritten contracts may be created even when unintended, simply as a result of what one or both of the parties do or say, and may result in equally unintended consequences.

Make every effort to include your entire agreement in a written document. Once a contract is executed, discussions you had beforehand, e.g., at a pre-bid conference, are generally not enforceable if not written in the contract. Look at any proposed agreement as a whole. The apparent meaning of a sentence you focus on in the agreement may be changed by another sentence in that same agreement.

When you think you need legal expertise in dealing with contracts, hire your own lawyer—not one who represents both you and the owner. Consulting your lawyer for advice before plunging into unfamiliar territory is much less costly in terms of both time and money than hiring him to get you out of the muck later on. This is especially important during your early years in the business. Through experience you will learn about various contract issues, and the occasions you need to call on your attorney will become fewer. Establish a relationship with a lawyer who specializes in construction law.

10.3   A few generalizations about contracts

According to my lawyers:

- Terms that have a special meaning should be defined in the contract.
- Specifications take precedence over the drawings.
- Handwritten contractual terms are given more weight than typewritten terms. Typewritten terms have priority over printed ones.
- If an ambiguity in the contract arises in the course of a dispute, the courts generally rule against the party who wrote the ambiguous language. It might seem that the contractor has the advantage here when the owner draws the contract, but if you don't bring unclear issues to the owner's attention and clarify them prior to bidding you may be responsible for any consequences of the ambiguity during construction.
- The owner has an implied duty to cooperate with the contractor, including coordinating the activities of suppliers, vendors, and others the owner has contracted with but who have no direct link with the contractor. Another implied duty of the owner is to provide the contractor with accurate and sufficient plans and specifications.

Conditions contrary to your interest can hide in plain view in the written language of a proposed agreement or may exist because of the absence of terms to the contrary.

10.4   Know the project owner

Anyone can form a corporation, and that corporation may or may not have assets that will be available for payment of its obligations. You may think you're dealing with a well-known, high-profile corporation only to learn when it's too late that the named "Owner" on your contract is a wholly owned but hollow subsidiary with a sound-alike name.

The contract must state the owner's name exactly as shown on his or his company's corporate registration, financial statements, and any other documentation you rely on. Your bank or bonding company can help you determine the finances and the payment practices of the "Owner" as named in the contract.

10.5 Getting paid

In a perfect world the owner hands you a trouble-free building site, develops a clear and accurate set of drawings and specifications, arranges financing including contingencies for change orders, offers you a contract that allocates the risk fairly between you and him, and pays promptly when you submit your applications for payments. All of your employees, subcontractors and suppliers perform as expected, and there are no delays on the project.

Sounds great, doesn't it? Of course, in the real world of construction you have to prepare for every one of those conditions and more to go *against* you, and the time to prepare for that is before you commit to the owner. Do not wait until after you sign the construction agreement, begin work, or are ready to apply for your first progress payment.

Even the best preparation does not guarantee that your applications for payment will be paid on a timely basis, but there are steps you can take to improve your odds. Be certain the construction agreement states

- When you may apply for progress payments
- The method for computing the amount due
- The documentation you're required to submit along with the application
- The amount of retainage the owner is allowed to withhold (the percentage preferably decreasing as the project progresses)
- Where, to whom, and by what means to send your application for payment
- When the owner is required to pay a properly submitted application for progress payments and final payment (i.e., the maximum number of days after he receives your application)

- The requirements for final payment application, which are usually different from progress payments

If the proposed contract requires you to submit lien waivers as a condition of payment, try to reach agreement with the owner that waivers for prior payments rather than the current payment will be acceptable. Paying all of your subcontractors and suppliers for amounts included in your current application for payment can cause you a logistical and especially a cash-flow problem.

If you're required to sign a lien waiver or claims waiver before you receive payment, state in your waiver that it is conditioned on your receipt of the payment it covers. Be certain that nothing in the owner–contractor agreement limits or prohibits your right to place a lien on the premises.

Here are a few non-contractual issues to explore in the interest of getting paid for the work you do.

- Where is the owner's money coming from? Bank financing? Cash in the bank? These are legitimate questions the owner should be willing to answer and document to your complete satisfaction.
- Will the owner's construction lender agree to set aside funds adequate to cover your contract amount plus reasonably estimated contingencies?
- What is the owner's track record for dealing with and paying contractors?
- Who in the owner's organization is authorized to approve change orders?
- What are the requirements for placing liens in the state and county the project is located in?
- If the owner's lender requires you to sign an agreement that you will complete the project in the event the lender takes it over due to default of the owner's obligations to the lender, does the lender agree to *first* pay you for any work done to date for which you have not been paid?

Some of these precautions and others are discussed in more depth elsewhere in this book.

10.6   Commencement/completion dates

The agreement should state the beginning date for the project, the number of days allowed for completion, and the provisions for time extensions for excusable delays. If the agreement provides for a Notice to Proceed to be issued by the owner or architect, it should state the number of days between the notice and the commencement of the construction clock.

10.7   Owner delay

When an owner causes a delay, the contractor is usually entitled to a time extension and compensation for his costs that result from the delay unless the contract language prohibits it. Therefore, astute owners try to include a "no-damages-for-delay" clause in the owner–contractor agreement, but you should negotiate this out of the contract as it unfairly places risk on you.

Even if a no-damages-for-delay clause is present in your agreement, you may be entitled to payment by the owner for delays caused by defective plans and specifications or his failure to supply labor or materials he's responsible for, make the site accessible, or pay you as agreed. These are known as compensable excusable delays.

Non-compensable excusable delays are delays for which the contractor is entitled to a time extension but not compensation. A common example is weather that cannot be reasonably anticipated, such as hurricanes and tornadoes. Labor strikes also fall into this category. You may be entitled to a time extension for them, but not payment for any economic loss unless the construction agreement states otherwise.

10.8   Contractor delay

In the event of late completion of a project, the owner may suffer consequential costs he would like to recover from the contractor. Consequential costs might include loss of the revenues the owner would have realized between the contractual completion date and the actual completion date, and additional interest on loans. Delays caused by unavoidable events such as exceptionally bad weather, changes in the work, and acts of God are generally excusable unless otherwise provided by the owner–contractor agreement, but in the case of unexcused delays the contractor may be liable for the owner's actual damages. It is not hard to imagine that the owner's losses can quickly become large.

Owners and contractors often agree to a contractual liquidated damages provision whereby an amount the contractor will pay to the owner for unexcused delays, usually a specified amount for each day of delay, is stated in the agreement. There are a couple of things to consider in establishing the liquidated damages amount: The owner's circumstances must be such that actual damages would be hard to calculate, and the liquidated damages amount should be a reasonable estimate of what the owner's actual damages would be, as established at the time the agreement is executed. While a small liquidated damages value might seem advantageous to the contractor, it could leave you facing a project owner's actual costs in a dispute if they prove to be substantially greater than the amount stated in the agreement. An owner's "actual" costs determined in this atmosphere can become inflated, and difficult and expensive to disprove.

10.9   Changes in the work

In order to make design changes, to correct errors in the drawings, and for other valid reasons, project owners must have the right to require changes in the work after the owner–contractor agreement is signed, with appropriate adjustment in the contract amount and time.

You can strengthen your claims for payment for extras by taking the following into account before doing the extra work:

- The person ordering you to do the extra work must be duly authorized to do so. This may seem obvious, but it is frequently a matter of dispute. If the construction agreement does not name the person or persons having this authority, try to reach an understanding with the owner before or as soon as possible after the job starts, and confirm it in writing to him and inform your project manager and job superintendent and others in your company who need to know. You risk not getting paid for the work you perform without proper authorization.
- The need to perform extra work is often initiated by the contractor when his project manager or job superintendent comes upon a situation that doesn't

work according to plan, or the owner wants a change in the design. Change authorizations often happen via telephone or other informal communication when it is not practical to wait for a formal change order to be drafted and signed before doing the work. All too often, contractors proceed with the work after receiving approval by word of mouth and assume the formal change order will come. This is risky. If you make the judgment call to proceed on faith, first be sure the oral (verbal) "approval" comes from a duly authorized person, then follow through immediately to confirm the change in writing while the details are fresh in the minds of all involved. Do not use the convenient, open-ended formula for disaster we've all fallen for: "Don't worry about the cost right now, we'll work it out later."

- Delay in formalizing a change order sometimes means lost or forgotten changes that will never be billed, or time-eroded details that may result in disputes or down-negotiated extra costs. Other factors such as an owner's history of issuing and paying oral change orders and a contractor's history of accepting them from that owner can affect the way change orders are handled between contractor and owner.

- A change order must state the amount the contractor is due for the work or specify how the amount will be determined—usually a lump-sum amount agreed to at the time the change is approved, a unit price, or on the basis of time and materials cost plus contractor markup for overhead and profit. Procedures for documenting change order costs are further discussed in Differing conditions (Changed conditions) later in this chapter.

- To be entitled to payment, you must be able to show that the owner's change order amounted to a change in the contract documents and to prove how much it cost you. You must not have caused the situation that requires the extra work. Of course, change orders can cut both ways: The contract amount can be reduced by changes that reduce your cost.

- While owners are entitled to require changes, the changes must be within the general scope of the underlying contract. An owner may require you to increase the size of a building under contract (within reason), but he is not entitled to require you, for example, to construct a second building on another tract of land.

As a niche contractor working with many customers repetitively, I tried to minimize change orders by sometimes absorbing relatively small extra costs even when a claim was justified. A certain amount of give and take is expected in repetitive work situations. But I've learned that owners often become very exacting and guarded when it comes to change orders that result in substantial increased cost. I've seen owners pull out all the defensive stops even when reasonable in-progress verification procedures were followed. Effective prior planning and continuous communication increase the owner's confidence that your ultimate charges for the extra work are fair and reasonable, which is especially important for contractors who work with repeat customers.

10.10   Constructive change

A constructive change occurs when the owner requires the contractor to do work that is different from that required by the contract documents. Here are examples of constructive change:

- When the contractor performs work based on the owner's defective plans and specifications and must correct it
- When the owner denies the contractor a justifiable extension in the construction completion date
- When the owner requires the contractor to do extra work due to the owner's misinterpretation of the contract documents

The contractor may claim payment for work he performs under these circumstances, but constructive change is often a matter of owner–contractor dispute. This highlights the need for timely notice to the owner and thorough documentation of the circumstances that constitute the constructive change.

Project owners may try to shift the risk of changed conditions to the contractor by a provision in the bidding and contract documents. Be alert to this and don't accept it.

10.11   Differing conditions (Changed conditions)

Another important special case of changes in the work is that of differing conditions. A differing condition is a physical condition that is discovered in the course of construction that was not apparent or believed to exist or reasonably expected at the time of bidding. An example is the discovery of subsurface rock formations when digging building footings (although this might be disputed if, say, outcroppings of rock on adjoining property are visible from the jobsite even if not on the jobsite itself).

Some other common problems that may constitute changed conditions include soil that cannot be compacted to the specified density, below-ground man-made structures such as abandoned building foundations, and buried debris; also a concealed condition such as one that might exist in the wall cavities of an existing structure.

Your construction agreement should contain a clause that describes how differing conditions will be handled. Without this, you may find yourself eating the associated costs. In the absence of a differing conditions clause, insist on a blanket disclaimer of responsibility for concealed or unknown conditions in your bid or in the contract. You may have trouble making a case that a condition meets the concealed test if other bidders on the project detected and noted it to the owner, or if it was discussed in a pre-bid meeting that you did or did not attend.

Talk with your attorney even if a differing conditions clause is included in the agreement if you are not certain it gives you all the protection you need. Let me stress that a differing conditions clause is not a stand-alone safety net. You must thoroughly inspect the premises of a project at the time of bidding and report to the owner or architect any discrepancies between the planned work shown on the bid documents and the actual conditions observed on the site. This gives the owner the opportunity to alter his design or otherwise change course before making contractual, financial, and other possibly irreversible commitments to a project that may not be workable as planned.

10.11.1 What to do upon discovering differing conditions

If you discover differing conditions after committing to a project, there are steps you should take to protect or improve your chances of getting paid for any required changes.

■ First, immediately stop work on the portion of the project affected by the differing conditions and protect it from any further disturbance so that tests and evaluations can be made. (Do not take it on yourself to "just go ahead and fix" what appears to be a simple and insignificant problem, as you may find yourself bearing the full cost of correcting both the original problem and any direct and consequential conditions resulting from your "fix," the owner's argument probably being that you have denied his other alternatives.)

- Promptly notify the owner of the differing site condition by phone and in writing, making it clear in your notice that you believe differing site conditions exist. Do not assume that any meetings or discussions you've had about this situation fulfill the notice requirement.

- Thoroughly document the differing condition including date and time, description of the condition and how it was discovered, photos, witnesses, and a description of other circumstances such as weather, and any action taken by you, your employees, your subcontractors, and others before and after discovery. Of course, immediately begin to track and document the impact of the changed condition on your job cost in terms of dollars and time. Effective documentation and clear notice to the owner specifying differing conditions are of the utmost importance.

In some cases an owner may try to shield himself from responsibility for the accuracy of the contract documents, including drawings, specifications, subsurface data, etc., and saddle the contractor with the consequences of any errors or omissions. (Lawyers call this an "exculpatory" clause.) I would not accept such a provision. Note that if the contract does not contain an exculpatory clause in favor of the owner, the owner is responsible for the accuracy of the plans and specifications even if not so stated in the agreement.

As mentioned above, once you begin the work required by a changed condition or any other change order, maintain complete, detailed, and accurate cost records. Consider using separate files and accounting cost codes for change order work. In some cases it may make sense to arrange for continuous or periodic independent verification during the work process or require the owner to have a representative present during the work. Of course, judgment is required: Minor or straightforward change orders may not require such extensive procedures.

Your chances for satisfactory resolution of change order payment dim if you and the owner fail to follow effective procedures, beginning upon discovery of the underlying condition. It is up to you to steer this ship. The process should begin with a complete understanding of the problem by all involved parties, the proposed resolution, the desired outcome, the work process, the overall impact on the job schedule, and a clear statement of the method for computing the payment you will be entitled to for the extra work. This should be put in writing and signed by all parties involved at the start.

Just as the contractor has certain responsibilities in the event of differing conditions, the project owner does as well. Your claim for a change order may be strengthened if the owner has failed to tell you all he knows about a site, has misrepresented information, or if the plans and specifications contain inaccuracies that would not have been discovered upon reasonable visual inspection of the site.

If the differing/changed conditions clause in your contract is not clear as to its application to the situation at hand, if your contract does not include such a clause at all, or if the potential impact of an actual changed condition situation concerns you, discuss how to proceed with your lawyer.

10.12 Insurance

Ask your insurance agent to review any proposed owner contract to be certain your coverages meet all requirements for the project. Require the owner and all subcontractors to name you as "additional insured" on all their insurance policies related to the project, and obtain the respective policies and have your insurance agent examine them to be sure you are properly protected. Having insurance certificates that name you as "additional insured" is mandatory, but they alone do not explain your coverage.

Read more about insurance in Chapter 14.

10.13 Indemnification

Indemnification is an agreement that provides for one party to bear the costs for damages or losses of a second party. The contractor is almost always required to indemnify the owner to some extent.

Some indemnification clauses require the contractor to indemnify the owner *even if the owner is solely responsible for the loss.* Such a provision is unacceptable to the contractor. An acceptable form of indemnification requires the contractor to "indemnify and hold harmless" the owner against claims *caused jointly* by the contractor or the subcontractors and the owner.

In general, if you're required to "hold harmless" another party, you are responsible for any and all damages suffered by that party. You should contractually limit your exposure to the owner's claims for consequential or indirect damages.

Follow the general rule to not agree to any indemnification provision that cannot be covered by insurance. Require your subcontractors to indemnify you and the owner to at least the extent that you are required to indemnify the owner. When faced with any unfamiliar indemnification clause, have it looked at by your insurance agent and maybe even your construction attorney.

10.14   Warranty obligations

A construction agreement may contain express warranties that extend the contractor's obligations beyond a specific time period. Some warranties that do not specify a time limitation might not come into play until a defect is discovered, which could occur years after completion of the work. Such warranty provisions highlight the need to carefully examine the bid documents and contract documents and address issues of concern by limiting their effect through insurance, contractually, or by taking their potential cost into consideration in your bid and in the contract amount.

10.15   Limitation of liability

Owners and contractors alike may be responsible for "consequential" damages unless waived by the parties. Examples of consequential damages the contractor may suffer if not waived include the owner's loss of income, profits, financing, and reputation resulting from acts of the contractor. The possibilities are limited only by the damaged party's attorney's imagination. Insist on a waiver of consequential damages and consider forgoing the project if the owner will not agree. The AIA and AGC agreement forms include a waiver of consequential damages.

10.16   Governing law

The agreement should provide that it will be governed by laws of the state in which the project is located or in your state. Avoid a third state, such as where the owner's corporate headquarters are located.

10.17   Dispute resolution

According to *Smith, Currie & Hancock LLP's Common Sense Construction Law*, "The key to quick resolution of disputes is the use of systems designed to collect, preserve, and organize information, including documents, throughout the project." It should be your policy from the beginning to establish and maintain such systems. If you wait until you need them in a dispute it may be too late.

Disputes and disagreements are virtually a certainty in construction, but you can improve your chances of being dragged through litigation by implementing the following policies and procedures:

- Approach disputes with reason and logic rather than hostility and accusation
- Be selective in the project owners you work for
- Be thorough in employee selection, training, and retention
- Build a reputation of fair and honest dealing
- Build trust and goodwill among your customers, employees, subcontractors, and others you work with
- Include pertinent notes in your bid proposals when permitted, to avoid surprises during construction
- Establish and follow good documentation practices
- Have pre-construction discussions with project owners in which potential problems may be anticipated and dealt with up front
- Keep lines of communication open among all parties
- Recognize that it is sometimes better to swallow a bitter-tasting settlement pill than to be dragged through the legal system by the emotions that usually scream revenge
- Recognize where giving a little may buy a lot of goodwill; this is the alternative to winning an argument over a relatively insignificant change order and losing a customer

■ Screen subcontractors

■ Use capable cost-accounting systems and procedures

Litigation should be the last resort in any construction dispute. Although any disagreement can take a toll on the parties, none is as emotionally draining, expensive, and distracting from business as usual as preparing for a trial and the trial itself. Include alternatives to litigation in your contracts that must be exhausted before either party can force litigation. Both the AIA and the AGC standard owner/contractor agreement forms provide for alternative dispute resolution.

Contracts often include a provision that the owner may require the contractor to proceed with the work in the event of a dispute. Decline this if you can in order to keep your options open.

10.18 Contract termination by the owner

The owner may terminate your construction agreement for default and, if provided for in your contract, for his "convenience." Be certain the contract defines events that constitute default, establishes notice provisions, and provides reasonable time for you to cure the default. If the owner has the right to terminate for convenience, the contract should spell out the payment you will be entitled to, including whether you will receive anticipated overhead and profit (as if the job had progressed to completion). Your minimum requirement should be that you are "made whole," i.e., returned to your circumstances that existed prior to the project. Your construction attorney will suggest the appropriate terms.

Chapter 11

You and your employees

For me, intensity covers a lot of sins. If there's one characteristic all winners share, it's that they care more than anyone else. No detail is too small to sweat or too large to dream. Over the years, I've always looked for this characteristic in the leaders we selected. It doesn't mean loud or flamboyant. It's something that comes from deep inside.

– Jack Welch, *Winning*

In the early days of my construction business I hired a job superintendent whom I'll call Roy who clearly knew more about building a building than I did, but it soon became obvious that he had a couple of personal traits that affected his interaction with others he came into contact with, including my customers and subcontractors. I met with Roy to discuss this with him, thinking that making him aware was all that would be necessary. His response was such that I knew I needed to say more: Changing his ways was a condition of his employment. The crusty, burly construction man stared steely-eyed at this relative greenhorn contractor for seconds that dragged on like hours before Roy nodded slightly. He continued working for me for a while but things between us were never quite the same.

The need for thorough hiring practices was driven home to me by that event. Better interviewing and reference checks might have exposed Roy's habits and given me the chance to look into them before hiring him when either of us could have walked away, no harm done.

No single factor standing alone can guarantee your company's success, but your failure to hire and retain the right people can by itself assure your downfall. Organizations drift down toward the lowest common denominator, so hiring to high standards is essential if your goal is excellence.

It's easy enough to grasp that concept but the problem is knowing who the "right" people are, how to attract them, and then how to keep them happy and productive and build loyalty in them. Here are some of the important factors.

11.1 Who are the "right" people?

Beyond covering your job-specific needs, here are some important characteristics to look for in a prospective employee:

- Ability to think outside the box
- Enthusiasm
- High moral and ethical standards
- Leadership qualities
- Likeability
- People savvy
- Positive attitude
- Quality consciousness
- Result oriented
- Self-confidence
- Self-control
- Solid sense of responsibility
- Strong work ethic
- Team player

Southwest Airlines is often named as one of America's most successful airlines. The company's high-profile leader Herb Kelleher was quoted in *The HR Manager* after being asked how he maintained his company's culture. He answered,

> **Well, first of all, it starts with hiring. We are zealous about hiring. We are looking for a particular type of person, regardless of which job category it is. We are looking for attitudes that are positive and for people who can lend themselves to causes. We want folks who have a good sense of humor and people who are interested in performing as a team and take joy in**

team results instead of individual accomplishments. If you start with the type of person you want to hire, presumably you can build a work force that is prepared for the culture you desire.

Of course some hires have more long-range implications than others. Non-career and temporary employees are hired by your superintendent or project manager or bookkeeper with less fuss to a degree depending on the position. This chapter deals with hiring people to responsibly carry out key functions as your company grows and to grow with it. These are the people you want to have your personal stamp on and you will learn here how to do it.

11.2 Hiring the "right" people

Someone has said, "I'd rather have an A team with a B idea than a B team with an A idea." I agree.

Opportunities exist in construction for even the smallest of contractors. In the beginning you may wear many hats simultaneously—general manager, salesman, estimator, project manager, expediter, job superintendent, bookkeeper, bill collector, even office janitor—spending long days on the jobsites and evenings and weekends doing mental or office work. This may be unavoidable in your early days but unless you climb to the next level, freeing yourself to see and manage the whole picture while building up a staff to take responsibility for at least some of those functions, you're likely to stay in the bid–build–bid–build cycle in which you have no revenue during the bid phase and little time to pursue and bid on new work during the build phase.

You will not thrive in this mode indefinitely because of other demands that you cannot ignore: costs and billings; payroll taxes and quarterly income tax filings; hiring and benefits; insurance and audits; vehicles; office/storage space; supplier accounts and lines of credit; performance and payment bonds; complex contract provisions; government reports; cost controls; filing/retrieval systems; contractor licenses; computers and software; phone systems; marketing the company and acquiring new business opportunities; implementing policies and procedures; and so on.

If you're personally handling most of those responsibilities, there's not much time left for the productive side of the business. Faced with this reality, contractors often conduct cursory job interviews and don't take the time to check references, making hiring decisions mostly on the basis of first impressions they get in a hurried interview. That usually results in poor or mediocre performance, internal and external personality clashes, and high employee turnover.

Good hiring practices are more likely to result in acquiring people who not only can excel in your current need, but can grow with you and take responsibility for critical parts of the company as it climbs to new levels. When you read biographies of successful business people, you can see that many of them credit their success to their ability to attract and retain high quality employees for key positions.

11.3 Good hiring practices

Make these ideas part of your hiring process.

- Develop a detailed description of the position to be filled.
- Seek to hire the best employees available who meet your needs. It's short-sighted to make hiring decisions solely on the basis of cost. As with most things in life, price goes up with quality, and if your hiring threshold is mediocrity, your performance is limited to mediocrity.
- Place professional looking, well-written ads in local newspapers and online.
- Develop written materials and a Web site that present your firm in a positive, attractive light. Consider using a specialist for this if possible.
- Set up a screening process to quickly eliminate applicants that do not meet your basic requirements.
- Avoid the applicant who has changed jobs every year or two in a lateral move if you're looking for a long-term employee.
- Interview promising candidates in person, and allow them to express themselves. Ask open-ended questions and questions that relate to the position you're filling.
- Obtain releases from the candidate that allow you to check his history with regard to education, former employment, criminal and driving records, and statements he makes in his application.

- When checking references, assume they are reluctant to say anything negative about the applicant. Focus on the candidate's specific behaviors at that employer, not general descriptions. Ask about his role, responsibilities, and specific accomplishments as measured against expectations. Listen carefully for what the reference may be withholding and probe into it.

- Be aware that it is illegal to discriminate on the basis of race, color, religion, sex, national origin, age over forty, and disabilities. Various other federal and state laws and regulations may apply and only an employment specialist is likely to know all of them.

- Check references for the several most recent years.

- Applicants whose experience in construction has no relation to your needs may not be qualified for your job. A project manager or superintendent who has been building high-rises or oil refineries won't be at home on a $1 million project that requires a lot of simultaneous activities in a small space in a short time period.

- Hire for a probationary period during which you or the employee may end the relationship with no questions asked.

- You can decide to develop tests to assess the candidate's knowledge and skills that pertain to a particular job. His responses to a simple test can tell you a lot about an applicant.

11.4 The interview

Here are some questions to ask and observations to make during an interview.

- In preparing for the interview write down what you want to have learned about the applicant after the meeting is over.

- In listening to his description of prior jobs, try to learn whether the cultures at those firms are compatible with yours.

- Ask what his earlier co-workers would say about him if asked.

- Probe into his possibly unstated reasons for leaving his last job.

- Ask what he liked most and least about his last job.

- Did he try to make improvements/changes in his previous job? How were they received by the company?

- Ask him to describe a situation in his previous job for which he was praised. Criticized.

- Describe a work situation in your company and ask how he would handle it.
- Ask him to describe the duties and responsibilities involved in the job he's applying for so you'll know whether he understands it.
- Ask him to describe his technical skills as they apply to the job he's applying for.
- Ask him to describe his greatest strength. Weakness.
- Note the candidate's mention of people he has worked with who are not named on his application and ask for his permission to contact them. There could be a reason the candidate didn't list them as a reference.
- Mention that you routinely check references. Ask if there's anything he may have forgotten to tell you that could show up.

Additional questions you can use in prospective-employee interviews are included in Appendix 4.

If your interview and follow-up investigation don't clearly yield the candidate you're completely satisfied with, a second look at your other top candidates may be in order. But remember that you're looking for a certain set of predetermined characteristics along with the "right" gut feeling, not just the best that the current group of applicants has to offer. If you hold out for the right person you'll find him.

11.5   New employee orientation

There may not be a time when your new employee will be more interested in what you want him to know about you and your company than when you meet to welcome him on board, probably on his first day at work. An orientation meeting is not so much about what his duties and responsibilities will be—you'd better have covered that earlier in the hiring process. It's not about general company policies and procedures and benefits; that's in the employee handbook discussed below. And the details of his hiring package like salary and bonuses and any non-disclosure and non-compete agreements you've agreed on should have already been given to him in writing.

So your purpose in the orientation meeting is to get into matters beyond hiring issues that you want to put your own personal touch on. Things like your values, your dreams for the company, the things that can be. The impossible. You want this person to catch your fire if he's one who can. This is a heart talk, not a speech. I've had employees remind me of something we discussed on their first day many years later, as if it was etched in their minds.

Only time will tell if this applicant is one of those rare ones. There will be many you just knew were but who surprised you. Not that they don't produce: They steadily and reliably get the job done and you're thankful that you have them. Then that one-in-a-hundred guy comes along who can make the hair on your arms stand up but you may not know it until later. Money can't buy what he has and it can't be taught. He just has it and he's putting it to work for you along with more loyalty and dedication than you could ever hope for. He's driven internally. You don't need to mention office hours or being at meetings on time or working late some nights or coming in on a Saturday when things get stacked up. All he needs is to be given his responsibilities and authorities and pointed in the right direction. He's hard-wired to do all the rest.

If your firm is not a place where he can be fulfilled in the course of carrying out his responsibilities to you, he'll find one where he can. He wouldn't work harder if he owned the company. Out of hundreds of fine employees who've graced my company over the years, I had more than a fair share of employees like that, both men and women, and for that I feel fortunate and grateful.

11.6 Non-compete non-disclose agreements

Some of your employees will be exposed to your trade secrets, customer lists, subcontractor/vendor lists, and other information that your competitors would like to put to use for themselves. An employee may also develop a close working relationship with some of your customers who might follow the employee to a future employer, or he might use the customer to get his own new construction business off the ground.

Employers in construction and other industries and professions have tried to design agreements that prevent this kind of employee activity but they don't always hold up in court.

Possibly more effective than a written agreement is the development of mutually trusting and loyal relationships with your employees and your customers, as well as others in your business sphere.

I've never had to try to legally enforce such an agreement with an employee, as none dishonored the letter or the spirit of the agreement he had signed. Nevertheless, I recommend using non-compete non-disclose agreements for certain key employees. Sometimes they hold up, at least in part. If nothing else, they record what was considered a fair bargain by both parties at the time they were signed, and remind honorable men and women of commitments they made in good faith.

There is no standard form for such agreements. Your best chance at having an enforceable restrictive agreement is to hire an attorney who specializes in employment law in the state or states you do business in. He knows applicable statutory law and recent court rulings that apply in your situation, which are in constant change.

11.7   Managing employees for the long term

Hiring the right employees is a crucial step toward business success, but now you must manage them effectively. A 2004 survey by the Corporate Executive Board, *Driving Performance and Retention Through Employee Engagement,* finds that only 11 percent of 50,000 employees in 59 organizations surveyed constantly try to improve, while at the other end of the spectrum 13 percent exhibit little commitment at all, often give minimal effort, and have a high turnover rate.

I understand these survey results to mean that if you've done a good job selecting an employee, his near- and long-term value to you then depends on his commitment to you and your company. I believe such commitment depends on factors that are largely within your control. Some of these factors are described below.

11.7.1   Relationships

It goes without saying that it is essential to maintain an effective relationship with your employees, and you do this in almost every conceivable way. In a small firm, you know the husbands, wives, and children, and attend the weddings, wakes, and funerals. Not only out of duty but because your employees and their families become your family and you care about their well-being and share in their lives. And as the provider of perhaps most or all of the income in any given employee family, you are seen by them as an authority figure that they necessarily depend on, and this imposes on you a certain role that you must fulfill.

Even so, you're the boss, and at times that requires the kind of tough coaching, discipline, or even firing that is difficult in any circumstance, but is near impossible when you and an employee have gone beyond the employer/employee relationship to, say, the point of becoming drinking buddies who have no reserves between you. General George Washington, quoted by historian David McCullough in his book *1776*, understood this pitfall, telling his Continental Army officers with regard to their relationships with the men under them, "Be easy . . . but not too familiar, lest you subject yourself to a want of that respect, which is necessary to support a proper command."

The deeper and longer this kind of fraternity grows, the harder it will be for you as the owner of your business to revert to the superior position if that time comes. Many a boss, so trapped, has delayed or neglected to level with an employee who, perhaps despite his best efforts, is not right for the job, instead of telling him up-front that he's not meeting expectations and that his future there is not what you and he would like it to be. Except in extenuating circumstances, that doesn't automatically mean you'll have to fire him. Faced with reality, he may make the decision to leave on his own.

Failure to deliver such bad news is almost never in the best interest of the employee, who is allowed to go on unaware, or unwilling to recognize, that he has little or no hope for advancement and is thus delayed in finding work for which he may be better suited and could excel in; or of the employer, who can't fill his position with someone who might be ideal for his job.

To be fair, I must say I've known successful employers who do not subscribe to this relationship principle and it may not be a problem for them. It just seems to me that it's better all around to let an employee know where he stands—good or not so good—on a somewhat regular frequency, and that is difficult if not ineffective if the propriety of the employer/employee relationship has taken a back seat to the personal relationship. This is a matter of personal management style, but you should weigh the consequences in deciding what is best for you.

11.7.2 Autonomy

Some experts believe that an employee's latitude to decide how he will accomplish the requirements of his position is the most important factor influencing his job satisfaction. Of course you must clearly establish parameters and objectives, but within that framework allow a capable employee to call his own plays.

11.7.3 Recognition

If the fruits of our labor as human beings are noticed and praised, we get a satisfying feeling that we want to perpetuate. One job you have as a leader is to recognize and nurture desired behaviors in your employees—not just once a year in a performance review but when you visit his jobsite or meet with him in your office about something. Send him a note about an impossible schedule he managed to meet or ask him about his daughter's recent graduation.

11.7.4 Employee's return on investment

Just as you've foregone other opportunities in order to own your own business, your employee is investing in you by giving up his opportunity to apply his abilities elsewhere, and needs to feel he has made the right choice based on his total return, which includes not only direct compensation but also his benefits, opportunity for advancement, and general satisfaction.

Be sure the salaries and, where applicable, bonuses you pay are competitive in your market. You can increase compensation without locking yourself into higher salaries through creative bonus plans that are based on objective, measurable criteria for one-time events. Avoid bonus or incentive plans that require you to make value judgments, which hold great potential for creating ill will among employees.

11.7.5 Employee incentive plans

Many contractors pay bonuses/incentives as part of their employee compensation package. *Merriam-Webster's Collegiate Dictionary*, Tenth Edition, defines bonus as "something in addition to what is expected or strictly due." Employees too often forget the essence of that definition and, unless the plan is properly designed and carried out, the "bonus" becomes expected. Once the employee begins to count the bonus as part of his base pay, the power of the bonus to influence realization of the desired goal and its usefulness as a reward are lost. Additionally, this can lead to dissatisfaction or worse in both employees and employers.

You can avoid such plan failure if you will first define for yourself the ultimate goal of the bonus in objective terms. If a bonus plan is to be based on some parameter of performance of your company as a whole, that parameter must be unambiguous. One problem with this kind of bonus plan is that every employee covered under the plan is rewarded irrespective of his individual performance. An advantage is that it tends to promote team effort.

You might prefer a bonus plan that is geared to individual results. In this case, you must understand what performance by an individual employee results in the ultimate objective of the bonus and must design a plan that is most likely to result in that performance. This can be tailored to various individuals or departments within the company. A potential drawback of this kind of plan is that one or more employees might earn a bonus during a period when your company as a whole made no profit; those bonus checks are painful to cut. If you rule such a possibility out of your incentive plan, the plan description must make that clear.

One example of an individualized plan might be used to control the general conditions cost on a project. You would offer to pay a portion, perhaps half, of any savings from the project's independently established general conditions budget to the employee(s) who has the authority and responsibility to control it (project manager, jobsite superintendent, or both). General conditions cost can be a volatile variable and easily go unmanaged in the heat of a project, and such a plan may result in a greater focus on this area that is often ripe for neglect.

The above examples are common, but bonus/incentive plans can be devised to address many different sets of circumstances.

An incentive plan that is most likely to result in a win for both the employer and the employee includes the following basics:

- A clear management objective
- Management's identification of exactly what the employee must do to accomplish that objective
- The employee's clear understanding of what criteria entitle him to a bonus, as well as factors outside his control that may affect it
- Good faith intentions by both parties
- Accurate and honest administration

11.7.6 Benefits packages

While direct compensation is important, employees now value the indirect benefits more than in the past. The benefits package you offer is a major influence in attracting and retaining valuable employees. Popular benefits include health, life, and disability insurance plans; time-off programs; retirement/savings plans including profit sharing, 401(k), Flexible Spending Accounts, Simplified Employee Pension (SEP), and one that is known as the SIMPLE plan. The various plans are strictly regulated and you will likely be better off to hire an outside benefits administrator to help you understand them and their impact on your finances, choose the ones best suited to you and your employees, and administer the plans going forward. Those who provide administrative services for a fee include insurance companies, consulting firms, payroll service companies, banks, and investment brokers.

Being the sponsor of your benefits plans places serious responsibilities on your firm. You may contract some of them out to an administrator but you will continue to be ultimately responsible. Of course, your administrator will be held responsible as well.

11.7.7 Trust

Your employee must be able to believe that what you say is true. A broken promise, even if it's unavoidable, puts a chink in that trust. Promise only what you intend to do and strongly believe you can do. Your "maybes" are heard by your employee as promises, so under promise and overproduce. Expect the same level of honesty from your employees.

11.7.8 Work/life balance

Ability to spend reasonable time with family is another important factor in an employee's job satisfaction and loyalty. Build in flexibility to accommodate children's special events, family illnesses, and other life situations. Earned vacations should be taken.

11.7.9 Work fulfillment

This is the employee's sense at the end of the workday that he has accomplished something worthwhile, maybe even something no one else could do as well, and that he has had the opportunity to use his talents and abilities to his fulfillment.

11.7.10 Training

Periodic and ongoing training programs allow the employee to improve and to learn new things he can put to work. This increases his value to the company and to himself. When you commit the time and money to an employee's training, he sees it as an indication of the value and confidence you place in him. Training may be provided in-house or through industry organizations such as the AGC, ABC, industry consultant FMI Corporation, and others.

An effective training practice for your employees is to make presentations to each other periodically. A review of the management and operational strategies and tactics which have worked and not worked for the presenting employees should be among the topics of discussion.

Standardize work procedures within your company to make it easy for employees to understand and follow them. This reduces mistakes and improves quality of work, customer service, and cost.

11.7.11 Job security

While there really is no such thing as guaranteed job security for an employee, your integrity, company history, business philosophy, ethics, and handling of terminations and other employee situations in the past are some of the factors that influence how your employee evaluates the security of his employment. If he sees these in a positive light, he will be more confident about his job even in the inevitable down cycles.

Mistakes happen. An employee needs to know there will not be a rush to judgment when he makes one. Weigh all the circumstances and judge him partly by how he responds to his error, and use his mistake as an opportunity for training. Of course, certain kinds of problems must be dealt with immediately (but not irrationally).

11.7.12 Internal conflict

Some internal conflict is probably inevitable, and it must be handled the right way. You may decide to meet with an employee and his manager together who are involved in some disagreement, but do so only as a last resort if they reach an impasse in their private discussions. If still unresolved, you can then meet separately with the manager and suggest or, if necessary, dictate solutions or compromises that he can present *as his own* to his employee. Your managers should follow this policy with the employees who report to them as well.

11.7.13 Openness and communication

No firm can open all of its records to its employees, but you should have as few business secrets as practical. Certain financial data and other strategic information are needed by key employees to do their work, but be sure they understand that they are to keep it confidential, and why that's important. Disclosure of the company's secrets can have dire consequences, so consider making confidentiality a condition of employment. Make your values known to everyone in the firm.

Be accessible. Invite employee input and give it due consideration. But no one should come to you expecting you to override something already put in place by his direct manager. Once you do you've diminished the manager's authority and created a degree of ill will all around. And that employee then has reason to think he can come to you for decisions any time he disagrees with his direct boss. Get bad news out of the way all at once if possible, but sometimes it's good to dribble good news out over time.

11.8 Responsibility vs job description

While a thorough job description helps a new hire become familiar with his duties and responsibilities quickly, it is only a guide. Consider a brain surgeon, who studies a patient's MRIs and other diagnostics prior to surgery in the course of planning what he must do in the operating room to remove an invasive tumor. Later on, during surgery, the doctor's responsibility is not to blindly follow the plan he made from all the data he had at that time, but to recognize and work around the actual conditions he runs into in order to best deal with the tumor. Similarly, your employees must focus not on job descriptions but on outcomes.

Your bookkeeper's job description may specify a list of procedures to follow in the course of providing job cost reports, and by following those procedures the bookkeeper may satisfy the job description requirement. But unless he "owns" the data in the resulting reports he has not taken responsibility for outcomes. When you're mulling over your P&L statement for the current month, you don't want to hear your bookkeeper say he followed all the rules in generating the report; you want him to confidently tell you it is accurate. Period. Rules and job descriptions are no substitute for human initiative. Clerks follow rules. People who are given responsibility and authority are accountable for outcomes.

The concept of exercising authority is similar. Be sure your job superintendents, project managers, and other supervisory employees internalize and effectively exercise the authority they're given. For example, in dealing with a headstrong subcontractor, your project manager may resort to using you as a crutch: "Joe, my boss needs you to get this problem fixed by Thursday," which makes the project manager look weak. It may result in Joe complying, but your project manager just gave up his authority to effectively manage Joe, especially when Joe is faced with a tough requirement, in which case he will demand to meet with you for a final decision. Your project manager executes the authority you have given him by stating his instructions to Joe as his own—not the boss's: "Joe, here's how I want you to do this." If the subcontractor does come to you and gets nowhere, you have reinforced your employee, and Joe no longer has any illusion about who owns the authority to direct him.

An employee in a position of responsibility who does not grasp and implement the responsibility and authority concepts cannot be depended on or will be ineffective; he should not occupy any of your key positions.

11.9   Evaluating employee performance

In a small firm, you may or may not elect to conduct formal periodic employee reviews. You're likely to be around your employees a lot of the time to observe them firsthand as they perform their duties. To avoid commenting on performance too often, keep a file of minor things that come to mind that you want to mention and wait until the right opportunity presents itself. Address special situations at any time.

Consider the following as criteria for employee evaluation, whether or not you conduct formal reviews. Modify them to meet your particular situation:

- Attitude
- Budget control
- Confidence and competence in managing others
- Dependability
- Estimating accuracy
- Follow-through
- Organization
- Persistence
- Relationships and communication within the company and with subcontractors, customers, suppliers, etc.
- Results—not activity—orientation
- Scheduling

- Self-control
- Sense of responsibility
- Sense of urgency

Lying beyond your routine coaching and direction of your employees is the need to clearly communicate and document an employee's failure to meet the requirements you've established for him in his position. You must explain to him the specific deviations and what he must do to correct them, and give him regular feedback on his progress. Document all of your meetings and discussions as they occur. Following these procedures may put your employee on the track for satisfactory performance, but if not, and you fire him, you will be in a better position to defend yourself against any legal action that may result. Your best bet is to get "preventive" guidelines from an employment attorney, which is infinitely more cost- and time-effective than fighting even the most frivolous employee lawsuit.

11.10   Employee termination

Just about the hardest job you will ever have to do is fire someone. It goes to the innermost part of the terminated employee's psyche as well as to many other aspects of his life. I've had to look away as grown men cried, and on one occasion a terminated superintendent's wife called me two weeks after the fact and demanded to know why I had not been paying her husband's salary: It turned out that he had not been able to tell her he'd been fired (due to an alcohol problem). It is almost impossible to be indifferent about letting someone go even for cause, but there are times when termination is unavoidable due to job performance, conduct, or economic reasons.

Even though most states have employment-at-will laws which give the employer the right to fire an employee for any reason or for no reason, courts and juries may side with the terminated employee in a lawsuit against his employer. Consult an attorney with experience in employment law in your state when there is a threat of legal action by an employee.

Once you've decided that termination of an employee is inevitable, do not delay. If the employee happens to get hurt on the job or becomes pregnant or files any sort of lawsuit against you, termination of that employee becomes impossible, at least for some time. Above all, avoid spur-of-the-moment firings.

11.10.1   Conducting the termination meeting

After you've done your homework and made the firm decision to terminate an employee, follow these guidelines:

- Meet with the employee where you can speak privately without interruption.
- Factually state the reasons for your decision.
- Say clearly and without rancor that you're letting him go and that your decision is not negotiable. Don't be dragged into an argument or cloud the issue with blame and recrimination at this point. Your objective is to end this on as positive a note as possible.
- Be professional and stay above the fray.
- Explain vacation pay, available continuation of health insurance through COBRA[1], any severance pay, etc. Put everything he needs to remember in writing.
- Go over the terms of any non-compete or non-disclosure agreements in place and tell him you expect that he will abide by their terms. Give him a copy of the agreements.
- If you are offering severance compensation beyond the usual in exchange for the employee's release of any claims, go over the terms of the release and have him sign it.
- Give the employee a note listing everything you expect him to do before leaving, such as turning files over to another employee, settling expense and other company accounts, and returning all company property, records, lists, keys, credit cards, company identity card, etc.
- Tell him the effective date of termination. If it is immediate, have his final check ready to give to him, subject to anything you require him to do before receiving it. In my experience, it is often better for all concerned to make termination effective immediately.
- Deal with any unfulfilled commitments you had made to the employee during the course of his employment.

---

[1] US Congress's "Consolidated Omnibus Budget Reconciliation Act of 1985."

- Mention that you expect lines of communication to remain open between him and the company to facilitate any outstanding matters.
- When terminating an employee that you have reason to believe may take legal action against you, discuss in advance your termination plans with an employment attorney. It is probably a good idea to have another manager present at the termination meeting.
- At the end of the meeting, offer your hand and wish the person well.

11.11 Employee handbook

The purpose of the employee handbook is to tell employees important things they need to know about their employment relationship with the company. You may think an employee handbook is overkill for your small company, but you'll have to think through policies, rules, and regulations at one time or another—with or without a handbook—so why not just hammer them out and formalize them right from the start. They do take time, but generic policy handbooks available at bookstores and online may serve as a template you can tailor to your liking. An employee handbook adds a level of professionalism to your company in the eyes of your employees.

Here are many of the issues that should be addressed in your employee handbook.

- A statement that the employee handbook is the guide to the employee/company work relationship
- Absenteeism and late reporting rules
- Accident reporting
- Avoidance of conduct that reflects negatively on the company
- Company benefit programs, including financial plans
- Company vehicle rules, if applicable
- Confidentiality of company business matters
- Confidentiality of salary and bonuses, if that is your policy
- Customer/vendor/employee relationships
- Disclaimer in which it is stated that the company handbook does not constitute an employment contract; that the employment relationship is at will, and that the company has the right to revise its policies without notice

- Dress policy
- Generic duties and responsibilities applicable to all employees
- Emergency and personal time off
- Employee business expense policy (business meals, entertainment, travel, rental cars, company vehicle, credit cards, handling of receipts, etc.)
- Employee problem-solving procedures
- Employee/employee dating policy
- Employment of relatives
- Equal employment and non-discrimination
- Exempt vs non-exempt status with regard to overtime pay
- Leaves of absence
- Paid holidays
- Performance reviews, if applicable
- Personal phone calls
- Probationary period, if any
- Harassment policy
- Resignation notice
- Safety issues
- Smoking and tobacco policy
- Use of and accountability for company property
- Use of e-mail and the Internet via company facilities and in company time
- Use of personal vehicles for company business
- Vacation policy
- Workplace/jobsite alcohol and illegal substance policy

The employee handbook should also contain the following forms to be signed by the employee and kept in company files:

- Employee certification and authorization form by which the employee certifies that all the information he provided on the employment application and during the interview and hiring process is true and correct; acknowledges that he may be required to take drug tests; authorizes the company to investigate his representations; authorizes any former employers, schools, and individuals to provide information to the company about the employee; and acknowledges that any misrepresentations may result in termination.
- Receipt form by which the employee acknowledges receipt of a copy of the company handbook.

Before publication, have the employee handbook reviewed by an attorney who specializes in employment law in your state. Some parts of it may be subject to certain federal and state laws.

11.12   Professional employer organizations

A professional employer organization (PEO) is a firm that takes over a small company's human resources duties and becomes a co-employer to the company's employees. It manages the company's human resources administration and usually can provide the company's employees better benefits through group buying power. Use of a PEO may result in lower labor costs to the small business owner and allow him to concentrate more on his business. This arrangement should not interfere with your management of your employees.

However, being lumped in with the employees from other companies may throw you over number-of-employee thresholds, perhaps out of small-business classification and the related benefits, subjecting your company to additional federal laws and regulations. Be sure to learn all the implications through a neutral source before taking this route and talk to other business owners who have used PEOs including those you are considering for your firm.

Chapter 12

You and your subcontractors

Communication is the most important skill in life.

      – Stephen R. Covey, *The Seven Habits of Highly Effective People*

B ased on my observation over time, general contractors specializing in chain stores tend to develop ongoing relationships with subcontractors to a greater degree than do non-specialized generals and there's good reason for it: This work is often repetitive. A subcontractor who becomes proficient in his trade within one or more chain store brands you work for offers value in speed and price. Project managers and superintendents forge effective working relationships with repetitive subcontractors, reducing time-consuming hand-holding and the likelihood of disputes. The cost-reducing benefits of the learning curve come into play.

When a project hits a snag, you're more comfortable with a subcontractor who has proven himself to you in the past, and who likewise knows he can depend on you. And you may be able to go to a familiar subcontractor in a crucial bid situation and get his cooperation in pricing to win.

Even with tried and true subcontractors you have to help them keep their prices in line by requiring competitive subcontractor bids on most or all of your projects. But you should be able to develop several subcontractors for each trade. Promote an atmosphere of fairness and mutual respect in your subcontractor relationships.

12.1 Independent contractor or employee?

Chain store contractors, like most other general contractors, rarely perform work with their own forces. It's better in most respects to outsource all of your jobsite needs except direct jobsite supervision.

At times it may not be clear whether a worker is an independent contractor or an employee for IRS classification purposes. Here are some of the tests the IRS says it may apply in determining the status of a worker. (The answers following the questions favor independent contractor status):

- Is there is a written agreement between you and the worker? (Yes)
- Do you provide employee benefits to the worker? (No)
- Do you control the worker's daily schedule, the tools he uses, and where he buys materials? (No)
- Who provides the tools and equipment needed to carry out the work? (Worker)
- Has the worker made a significant financial investment related to his work? (Yes)
- Do you reimburse the worker's expenses? (No)
- Does the worker have the opportunity for profit and loss? (Yes)
- Is the worker free to provide his services to others? (Yes)
- Does the worker have his own place of business and vehicles? (Yes)

These are guidelines the IRS may weigh against the situation in question. Of course, the agency will apply the above or any other tests it chooses.

The IRS gives the employer a strong incentive to correctly apply the classification of employee or independent contractor. If the agency determines that a worker you've classified as a contractor is by their rules an employee, you may be liable for certain costs, interest, and penalties. When you classify a worker as a subcontractor, the IRS requires you to report your payments to the agency on its form 1099-Misc, the equivalent of the form W-2 provided to employees.

Visit the IRS Web site at www.irs.gov for more information.

12.2   Subcontractor qualification checklist

Determine the subcontractors' length of time in business, suppliers and contact information, equipment owned, financial information, and a list of gerenal contractors you can call for reference. Check all references.

Once you determine a subcontractor's ability, it is up to your project manager and field manager to require him to perform at that level or better. A subcontractor may perform no better than is required or expected of him. Determine which of your subcontractors' foremen are the most effective on your projects and demand them on your future projects as a condition of the subcontractor doing your work.

12.3   The contractor–subcontractor agreement: Special considerations

The AIA and AGC publish subcontract forms, and many contractors develop their own. I favor the AGC family of forms, modified to your company's particular needs.

Use a subcontract agreement that addresses the spectrum of situations that may result in disputes. When a dispute arises you can yield on a tough provision in your subcontract, if necessary, to maintain fairness to the subcontractor; but in the absence of a contractual provision that would have protected you, you can't demand fair treatment from an unreasonable subcontractor.

Whether you use the AGC or AIA subcontract forms that are meant to cover all the bases, or a custom form that you create, below are issues that experience has taught me are often at the root of subcontractor disputes, and subcontract provisions that can help to avoid them. These do not address all elements of a complete subcontract agreement, and should not be considered legal advice. Have your construction attorney help you create or review any subcontract document you plan to use until you are completely familiar with its terms and conditions and its limitations of applicability.

Subcontracts should be made to conform with the prime contract between the contractor and the project owner, and must be revised in accordance with the changes made to the prime contract throughout the course of the project.

12.3.1 Pass-through or flow-down clause

This binds the subcontractor to the contractor as the contractor is bound to the owner. Without a pass-through clause, you may be liable for conditions imposed on you by the owner that you cannot require the subcontractor to follow. For example, the owner–contractor agreement, or prime contract, may require the contractor to "permit only skilled persons to perform the work." The subcontract pass-through clause imposes that requirement on the subcontractor.

Structure your pass-through clause so that in the event of termination of the prime contract by the owner, your payment obligations to the subcontractor are limited by the amount you are able to collect from the owner on the subcontractor's behalf. No direct contract exists between the owner and the contractor's subcontractors and vendors.

12.3.2 Scope of work

It is not sufficient to state in the subcontract that the subcontractor "shall perform all work in accordance with the plans and specifications." Each subcontract should include the following:

- The project owner's description of the subcontractor's work as written in the general contract
- Items of work included in or excluded from the subcontractor's scope of work when different from the plans and specs
- Clarifications as necessary to assign responsibility for items of work not clearly assigned to a designated trade
- Work that is necessary for a complete job but is not included in the plans and specifications
- Method for resolution of conflicts that may occur between different parts of the subcontract, i.e., agreement as to who will have authority to make definitive decisions that the contractor and subcontractor agree to submit to

12.3.3 Work as directed

Include subcontract language that requires the subcontractor to work as directed by you. This provides for change order work.

12.3.4 Changes to the subcontract

The subcontractor may be entitled to payment for his additional costs and/or a time extension when changes are made to the subcontract. The subcontract should specify which employees of the contractor are authorized to order changes and any limitations on amount. It should also name the subcontractor employees who have authority to commit the subcontractor. Include the method for computing the amount of overhead and profit the subcontractor is entitled to for authorized changes.

Include a provision that the subcontractor is not entitled to payment for change order work unless authorized in writing by the contractor's designated representative *before* the subcontractor does the work. Note that you may lose your ability to use this provision to deny payment in a particular instance if you have not enforced the provision consistently.

Memories fade with time, as do your chances of getting paid by the owner for extra work. The subcontract should specify, and you should enforce, a strict timetable and procedure by which the subcontractor submits the required documentation for change orders to you for your use in billing the owner.

12.3.5 Conditions for payment to subcontractor

Specify the documentation the subcontractor must provide and other preconditions for interim and final payments, and the time frames, forms to be used, and location to which they must be submitted to you. Require the following documents from subcontractors:

- **Subcontractor's lien waiver** You can require this for all payments including the current draw or, alternatively, for amounts the subcontractor has been paid prior to the current one.

- **Subcontractor's suppliers' and sub-subcontractors' lien waivers**
  Vendors to your subcontractor should be named by your subcontractor at
  the time the subcontract is signed or afterwards, and you should require your
  subcontractor to provide lien waivers from them. Also, there is nothing to
  prevent you from occasionally calling your subcontractor's vendors to confirm
  that he is paying them as agreed. All lien waivers should include the inclusive
  dates, identification of the person or firm waiving lien, and the name and
  location of the project they cover.

- **Subcontractor's affidavit** Require the subcontractor to provide a statement certifying that he has completed the work included in his current
  payment application in accordance with the contract documents; that all labor,
  materials and services have been paid for; and that he has complied with all
  federal, state, and local laws, including Social Security law, unemployment
  compensation laws, workers compensation laws, and has paid all taxes in so
  far as applicable.

- **Subcontractor's certificate of insurance** Have this in hand before the
  subcontractor starts work at the jobsite. Accept only original certificates signed
  by the subcontractor's insurance agency, and which name the contractor as
  "Holder" and provide for at least thirty days written notice to the contractor
  before the insurance may be cancelled. (Insurance is discussed in Chapter 14.)
  The notice provision is important. We once had a subcontractor on a project
  whose insurance was cancelled during the project, and we discovered too late
  that we had not required thirty-day notice prior to any cancellation. The standard certificate of insurance we received provided only that the subcontractor's
  insurance carrier "endeavor" to give us ten days notice prior to cancellation, which it did not provide. When one of that subcontractor's employees
  was seriously injured on our jobsite, our own worker's comp policy took the
  injury claim by default. This resulted in a large increase in our worker's comp
  insurance premiums for the next three years.

- **Other requirements as applicable to the circumstances.**

12.3.6 Pay-if-paid

Contractors prefer to structure subcontracts so that the contractor pays the
subcontractor *only if* the contractor receives payment from the owner, arguing
that they should not be stuck with paying bills that the owner doesn't pay them

for. Construction lawyers say that a contractual provision to this effect usually will not be upheld in court unless the pay-if-paid provision makes it crystal clear that it is the intent of the parties that the contractor is not obligated to pay the subcontractor if the owner does not pay the contractor.

If you intend to include a pay-if-paid provision in your subcontracts, it should be written as follows, according to *Smith, Currie & Hancock LLP's Common Sense Construction Law*:

> . . . **that the owner's payment is a 'condition precedent'; that the subcontractor expressly accepts the risk that the owner may not pay the contractor; that the subcontractor relies for payment on the credit and ability of the owner to pay; and [if applicable] that the contractor's payment bond surety will be obligated only to the same extent as the prime contractor.**

Before agreeing to such a provision, a sensible subcontractor is certain to factor what he considers the risk of non-payment into his price. Some courts have refused to uphold pay-if-paid clauses, and some states have laws that make them unenforceable.

12.3.7 Delay damages

The subcontract should limit any delay damages payable to the subcontractor as a result of owner delay to a percentage of the amount received by the contractor from the owner.

12.3.8 Retainage

Specify the percentage of retainage to be withheld from the subcontractor's payment. The amount of retainage the contractor is allowed to withhold may be regulated by law in some states.

12.3.9 Calculation of payment amount

Leave nothing open to interpretation in setting forth in the subcontract how the amount due to the subcontractor for progress payments is computed. A reasonable method is to add the value of labor, materials, and equipment incorporated into the project, plus the value of materials stored on site and paid for, plus the subcontractor's markup, and subtract the value of retainage on the current application and any prior payments made to the subcontractor. Determining percentage of completion at any point during construction is usually subjective and can be difficult. Your objective as the contractor is for the unpaid balance of the subcontract including retainage held at any point during the project to be sufficient for completion of the subcontractor's work.

12.3.10 Terms for final payment

Spell out when the subcontractor will be paid the final amount due. The following should be included in the conditions required prior to final payment to your subcontractor:

- Completion of all work included in the subcontract and any changes, including punch lists
- Written acceptance of the work by the owner (and architect, if applicable)
- Final payment by project owner to contractor
- Subcontractor's waiver of lien rights (note that a lien waiver benefits the property owner; the contractor is protected by the subcontractor's release of claims)
- Subcontractor's release of claims against contractor
- Subcontractor's submittal of all equipment warranties and operating manuals
- Subcontractor's submittal of as-built drawings
- Other requirements specific to the project as applicable

12.3.11 Indemnity

Indemnification is the concept under which one party guarantees another party against certain losses. Your subcontractor should be required to indemnify your firm, the owner, and your agents and representatives against damages caused partially or wholly by the subcontractor or anyone directly or indirectly working for him.

12.3.12 Termination for convenience

Provide that you have the right to terminate the subcontract without cause. In that case, the subcontractor may be entitled to compensation for work in place, paid-for materials stored at the jobsite, and overhead and profit earned to date. Without a subcontract provision to the contrary, the subcontractor may argue that he is entitled to overhead and profit as if the contract was fulfilled and to other compensation.

Even with a termination for convenience provision, you can't rule out having to respond to a lawsuit by the subcontractor.

12.3.13 Subcontractor default

The subcontract should specify certain circumstances that constitute default by the subcontractor and give the contractor the express right to terminate. These circumstances may include but not necessarily be limited to the subcontractor's:

- Failure to supply enough labor and materials
- Failure to pay for labor and materials as agreed
- Disregard for laws, ordinances, and regulations

In giving notice of termination or impending termination due to subcontractor default, use the word "default" and strictly follow the subcontract terms for giving any notice. If a surety has provided performance and/or payment bonds for the project, they probably obligate you to notify the surety. Notify the subcontractor's bank if you have agreed to do so. Your failure to give notice to a party as required gives that party a defense.

State in the subcontract that you are entitled to deduct your expenses, including attorney's fees, resulting from the subcontractor's default from the amount you would otherwise owe the subcontractor, and that the subcontractor is required to reimburse you if your expenses exceed the amount due to him.

12.3.14   Notice of default

Specify how notice is to be given by the contractor to the subcontractor in an event of subcontractor default.

12.3.15   Cure

Specify how the subcontractor may "cure" a material breach, and the amount of time he is allowed to do so after being given notice.

12.3.16   Contractor alternatives

Provide in the subcontract for various courses of action available to you if the subcontractor does not cure the default within the allotted time. These may include

- Bringing in a second subcontractor to supplement the subcontractor's work or to take over some part of the subcontractor's work
- Terminating the subcontract, including taking possession of the subcontractor's materials and equipment and completing the job with another subcontractor
- Advancing funds to the subcontractor for his use in continuing the project (this option involves obvious financial risk to you that you should weigh carefully before proceeding; it also may require notice to/approval by the subcontractor's bonding company, if any)

12.3.17   Continuation of performance

Provide that the subcontractor must continue to perform in accordance with the subcontract during the course of any dispute between contractor and subcontractor. You may be able to use this to prevent the subcontractor from pulling off the job to force your hand on a disputed issue.

12.3.18 Dispute resolution

Litigation is usually the most expensive, time-consuming, and distracting method for settling disputes. Consider including an Alternative Dispute Resolution (ADR) clause in the subcontract, by which various non-court avenues must be exhausted before either party may use the courts for settlement. Both AIA and AGC subcontract forms include ADR provisions.

12.3.19 Termination of subcontract

Contract termination can have extremely serious consequences and should be done in consultation with your attorney.

12.3.20 Merger

Include a merger clause, which provides that the subcontract agreement incorporates the entire understanding between the parties and may not be changed orally.

A well-constructed subcontract agreement is a key piece in the contractor–subcontractor relationship and contracting success, but no subcontract replaces the need for contractor diligence in documentation, scheduling, communication, supervision/monitoring, coordination, and effective relationship management. No matter the thoroughness of the written provisions in your subcontract agreement the subcontractor may read no further than the scope of work. It is up to you or your project manager to meet with a responsible employee of each subcontractor at the start of the project and clearly express the performance you require of him, including manpower and supervision levels, interim schedules, final schedule, quality of workmanship, change order procedure, safety, and other aspects of the subcontract agreement and the job that are of special importance to you.

Chapter 13

Banking and finance

You don't do business with an institution. You do business with people. When you get a banker who believes in you, you can accomplish incredible things.

– Debbi Fields, "The benefits of making your banker your friend"

M oney may be the root of all evil, as some say, but lack of it is undoubtedly the cause of many doomed construction projects, bankrupt contractors, and failed owners. Indeed, the US Small Business Administration (SBA) has said that shortage of capital is one of the main reasons for business failure, along with lack of proper management.

The steady flow of owner payment as agreed is the lifeblood of a construction project. If the project owner's source of financing doesn't deliver or if for any reason the contractor, subcontractors, or vendors of custom components for the project fail to receive payment for their work as agreed, one or more of those parties may be unwilling or unable to pay those below them in the money chain. Contractors and subcontractors usually have the contractual right to stop work when payment is overdue (even though that may not be the best course of action in a given situation). Custom-product vendors may refuse to ship. Lowe's and Home Depot credit lines can dry up. Lawsuits and business failures may follow.

The general contractor plays a central role in the money game. As discussed earlier, it is your responsibility to verify the owner's ability to fund the project, including a reasonable reserve for unknown costs. It also falls upon you to hire financially sound subcontractors and other vendors who have a history of completing projects on time and within budget. Their record is also an indication of the sub-subcontractors and suppliers they use.

This leaves you to make sure your own finances are in order. This chapter discusses some of the basic elements of contractor financing.

13.1   Your business plan

Entrepreneurs sometimes view preparation of a business plan as a burden to be endured only because lenders require it. Informed business people look at a business plan in another light. It is first and foremost a study of all the factors that will have an impact on the successful start-up, operation, and management of the company. If you're considering starting a new business venture, you'd be shortsighted to not dig out the relevant data even if you don't require outside financing. It will serve as your blueprint for growing a business in the future, and bring to your attention factors that might have gone unnoticed if you had not hammered out a business plan.

If you do need bank financing, a well-prepared business plan will convince your lenders that you've done your homework. A good Internet source for business plan ideas and templates is Business Owner's Toolkit, www.toolkit.cch.com.

13.2   Sources of financing

Your business plan projects how much money you need to get started, when you expect to become profitable, how you will repay any loans, and the amount you need in a line of credit to use when your cash is temporarily short. These projections will be necessary in establishing your banking relationships.

You may have accumulated enough assets to start and operate your business without borrowing money but if not, here are some of the practical sources of financing:

- **SBA**   This government agency has a number of loan programs for small business. The loans are made and administered by the participating banks and partially guaranteed by the SBA. Rates can be a bit higher than on standard loans but the SBA's backing allows banks to make loans to borrowers who would not qualify otherwise. SBA loan programs are worth looking into.

- **Community banks** Smaller, regional banks that generally offer personalized service. You can take your banker to breakfast or lunch now and then, and develop a congenial relationship that will serve you well. Community banks make long- and short-term loans that will likely suit your needs. Their loan limits are lower than the large banks' but this should not affect your construction business loan needs.

- **Mega banks** The banking industry is consolidating, making it harder to develop an ongoing relationship with a designated loan officer who will know who you are when you call.

- **Credit cards** Many start-up businesses are financed through credit cards. This method will not serve your needs well over the longer term, and interest rates can be very high after the typical "come-on" period is over.

- **Private lender** You may have a friend or business associate who is willing to lend you money. But doing business with friends risks the relationship altogether if things don't go as planned. Approach the loan strictly as a business transaction by having loan documents drawn up by a lawyer and signed by you and your lender. Abide by the terms of the agreement.

- **Family loan** Borrowing from a family member is similar to going to a friend for money, but with worse possible outcomes. A bad financial experience involving your family can have far-reaching consequences but if you must take this route, follow the procedures described for a private lender above.

- **Sale of investments** Your real estate, stocks, bonds, and other assets can be a source of financing without the payback burden.

- **Home equity** Banks offer a line of credit based on a percentage (60 to 80 percent) of your equity in your home, usually with reasonable repayment terms. The risk of losing the home you and your family live in is a strong reason to avoid this kind of loan. However, many other loan types require your personal guarantee as well, which may include your real estate assets. Home equity loan fees and rates are comparatively low.

- **Cash value of life insurance** You usually may borrow against any cash value in your life insurance policies with minimum payback consequences.

- **IRA or former employer retirement fund or 401(k)** These are good sources of funds with no repayment requirement. You may have to pay income tax and a penalty if you cash out of a plan.

13.3   Borrowing criteria

Lenders considering loan applications typically weigh these four factors (the four Cs).

- **Credit**   How well have you handled credit in the past, both personal and business? Evidence of the importance placed on your credit experience by lenders is the trend in retail credit to base lending decisions almost solely on the applicant's "credit score," as established by one or more of the national credit reporting agencies.
- **Character**   An evaluation of your standing in the community and how you've handled responsibilities and obligations in all areas of your life.
- **Capacity**   What backup systems do you have in place to rely on when things don't go as planned? Say the economy slumps and your customers quit building for a while. How will you survive? In making this evaluation, your lender will consider all of your assets, income, and expenses, and your contingency plan for bad times.
- **Collateral**   Your lender will probably require collateral to secure your loan. Qualifying assets include stocks and other securities, real estate, business equipment, and your firm's accounts receivable.

13.4   Managing credit

As indicated above, a good credit record is invaluable. To protect it, set up payment policies and procedures that result in paying bills when due. Paying as agreed is a strong incentive to your suppliers to increase your credit limits.

Be prudent in borrowing. Keep long-term borrowing in line with realistic earnings projections. If you have a line of credit, peg repayment of any drawdown to a known receivable that is due within thirty days. If you find yourself borrowing short-term money to cover operating expenses over longer periods of time, you're heading for trouble. Always know where the repayment money is coming from, and when.

When establishing a loan, request the loan documents ahead of time and have your business attorney review them before the closing. Lender's documents are one-sided in favor of the lender, but your lawyer may be able to negotiate more favorable terms with regard to some particularly onerous provisions. Be sure the loan documents require the lender to give you written notice of any default and a reasonable period of time to cure the default.

Most lenders and suppliers require you to personally guarantee loan repayment. You probably cannot avoid this with banks, but most vendors are less restrictive.

Chapter 14

Insurance and bonds

Nothing you will do as owner of your construction firm is more critical than securing the right insurance. This does not mean simply buying insurance policies. It means understanding what is and is not covered by those policies, where your exposure lies.

– Nick B. Ganaway

14.1 Insurance

Construction is hazardous and the risk of injury to individual workers and damage to property is significant. Comprehensive, single-policy coverage that meets the various insurance requirements of the owners, contractors, subcontractors, architects, engineers, and other parties to a construction project is available for only very large projects—generally those projects in excess of $100,000,000. For typical projects, the insurance industry offers a complex mix of policies that are intended to serve those parties' various needs to the extent that insurance coverage is available and economically feasible. Only an insurance professional with special expertise in construction insurance can competently analyze your particular needs on a given project, bring together the appropriate policies, and provide the continuing services you need in the course of your business—and do so at a competitive price.

An insurance policy is a contract between the carrier and you that spells out the carrier's obligations, your responsibilities, the benefits available to you, and policy exclusions. Your carrier will scrutinize any claim you make and deny coverage if all of the required conditions have not been met. Therefore the burden is on you to understand your coverage and to be diligent in fulfilling your responsibilities under your policies; your failure to do so could result in a possibly devastating uninsured loss.

The purpose of this chapter is to raise your awareness of the importance and complexity of insurance, mention the more common policies, highlight some of the pitfalls to watch out for, and possibly trigger questions and ideas to discuss with your insurance agent; it is not intended to be a comprehensive review of construction risk and insurance.

Here are only a few of the insurance policies you may need:

- **Property insurance**　Insurance on your own property, including offices, warehouses, and their contents
- **Commercial general liability**　Insurance that covers your legal liability for damages to property of others and injuries to individuals (other than employees)
- **Completed operations**　Insurance that covers you for liability from claims arising after a job is completed—a coverage that should be provided by the commercial general liability policy
- **Statutory worker's compensation**　Insurance required by law that covers claims by employees injured on the job
- **Automobile insurance**　Covers accident and liability coverage on your automobiles and other motor vehicles as well as damage to your own vehicle
- **Umbrella liability**　Insurance that will provide additional coverage for "underlying polices" (e.g., Commercial General Liability and Auto Liability)
- **Builder's risk**　Insurance that covers physical loss or damage to a construction project by an external source such as a fire or storm
- **Equipment floater**　Insurance that covers physical loss to your equipment regardless of location
- **Installation floater**　Insurance that covers loss or damage to machinery or equipment to be installed in the project
- **Transportation floater**　Insurance that covers property you own or are responsible for during transport
- **Business interruption or extra expense**　Insurance that covers your losses if your business is interrupted
- **Crime insurance**　Insurance that protects you from losses resulting from theft and other crimes
- **Employment practices liability**　Covers claims of discrimination, harassment, wrongful termination, and others

- **Environmental insurance** This coverage offers some protection if you contaminate the environment
- **Professional liability** Insurance that provides protection when you may be responsible for the design of the project or any component of the design

There are many other types of insurance and each has its share of critical details, inclusions, exclusions, deductibles, and policy limits, which you should discuss with your agent.

Following is a description of four of the insurance policies most common to construction contracts: Worker's Compensation, Employer's Liability, Builder's Risk, and Commercial General Liability.

14.1.1 Worker's compensation insurance

Each state has its own laws that regulate worker's comp (WC), but the underlying general principle is that employers are required by law to maintain statutory WC insurance for their employees and in exchange are relieved of direct liability for an injured worker's claims. A state's WC laws and regulations set forth the benefits that the insurance companies provide to the injured employee. The employer must buy WC insurance separately for each state his employees work in. Although most policies will automatically extend to provide coverage in each state, you should confirm that this is the case for your policy. Most states allow an employer to be self-insured in lieu of WC insurance coverage if the employer proves financial ability to pay all losses.

A WC claim goes through a prescribed process to determine the extent of the injury and its effect on the worker's employability, and consequently his ability to earn. The WC award may be temporary or permanent. In most cases, the award will not fully offset the worker's loss of income.

In some states, employers are not required to carry WC if they have fewer than a certain number of workers—often two or three. In most states, WC insurance is provided by private insurance companies in accordance with the respective state laws. A few states provide state-funded WC plans as the employer's sole option, and still others allow both state-funded and private insurance options.

14.1.2   Employer's liability insurance

Employer's liability insurance is intended to provide coverage for employee claims made outside of the state-mandated statutory system. An example of this situation is where an employee not only makes a WC claim for an injury sustained on your project but also makes a claim against one of your subcontractors. If the subcontractor in turn sues you, which is not unlikely, your employer's liability policy defends you and pays the amount you're liable for up to the face amount of the policy.

Another application of employer's liability coverage is where the employer fails to comply with state law with regard to working conditions and the injured employee chooses to sue the employer under common law.

Employer's liability insurance is written in conjunction with WC insurance except in states that provide their own WC programs. In those states, the contractor must buy employer's liability insurance by way of an endorsement to the Commercial General Liability policy or WC policy.

14.1.3   All-risk builder's risk insurance

Builder's risk insurance is widely used on construction projects to provide protection to the insured party or parties against physical damage or destruction of the project or temporary buildings on the jobsite by external causes such as fire and storm, and should include coverage for materials stored or in transit to the jobsite or waiting for incorporation into the project.

Builder's risk *does not* provide liability protection, or coverage for loss resulting from faulty workmanship or design errors. There are many other exclusions to the coverage that you should make yourself aware of, but some of the exclusions can be removed with a corresponding increase in the premium. A range of deductible amounts is available. The builder's risk policy may be written for a specific project, but the blanket form (with a reporting feature—i.e., monthly reporting) may be the better choice if you anticipate a steady flow of projects during the year.

The purpose of builder's risk insurance is to protect the interests of the project owner, general contractor, and subcontractors. Some policies require the cause of loss to be external. Lenders involved in the project may be protected. The owner–contractor contract usually specifies which party is required to provide the builder's risk policy, most often the contractor, who includes the cost in his price. All interested parties should be named in the policy as "additional insured" to assure the most complete coverage for all parties to the project.

Despite its limitations and exclusions, the builder's risk policy is flexible, and it is imperative that you and your insurance agent carefully match the provisions of the policy with your specific requirements and your needs in general. Although it is not economically or physically possible to cover every element of risk, I tend toward over-insuring in the interest of sleeping better at night.

Use care in specifying the coverage date of the policy to be certain that coverage is in place throughout the duration of the project. This requires diligence because the policy term may need to be changed if either the project start date or the completion date changes subsequent to the issuance of the policy. Also, builder's risk coverage usually ends upon occupancy of the project by the owner, which may occur before the project is complete.

In cases where the builder's risk policy is to be provided by the owner, you should request a copy of the policy and review it with your agent. If it does not cover all of your interests, you may be able to get the owner to have his coverage broadened, or to separately purchase the additional coverage you need.

14.1.4 Commercial general liability insurance

A Commercial General Liability (CGL) policy protects you against third-party liability claims for property damage or personal injury that arise from your own operations and acts, and the operations of your subcontractors or representatives (see "Note of caution" under Insurance administration, Section 14.1.6) and may be tailored to protect against other claims. Your contract with the project owner will specify the minimum dollar limits per occurrence and for all occurrences combined.

Your CGL policy may be tailored to protect you against the following kinds of claims, including providing legal defense, up to the policy limits:

- Injury or property damage to others that is caused by you, your employees, subcontractors, and others you are responsible for (see "Note of caution" under Insurance administration). Ask your agent about a severability or separation of interests clause which would permit the carrier to deny or limit coverage to, say, a drunken employee who creates a liability, but not to the company.
- Claims against you that arise out of a project after the work is complete (again, see "Note of caution" under Insurance administration).
- Legal liability that you accept by terms of your contract with the owner or others.

14.1.4.1 CGL endorsements

Your insurance agent can tailor your CGL policy to cover the above risks by way of endorsements to the basic policy. Here are some of the endorsements available:

- Per project aggregate limits (instead of per policy)
- Personal injury (libel, slander, defamation of character)
- Per occurrence coverage for bodily injury and property damage
- Additional insured (a party named as an "insured" on another party's policy)
- Premises-operations liability
- Independent contractor liability
- Completed operations and products liability
- Contractual liability including indemnity requirements in the owner–contractor contract
- Coverage for special operations such as underground excavation and blasting
- Host liquor liability
- Excess liability insurance ("umbrella") to provide liability coverage in excess of the limits of your basic policies

There are many standard exclusions to the CGL policy, some of which may be covered by your builder's risk or other policies. Be sure to know the areas in which you have no coverage.

14.1.4.2   CGL policy limits, deductibles, and legal costs

A CGL policy may be structured so that the carrier's cost of defending you is either in addition to or included in the policy limit. It is not hard to imagine that the legal fees and policy deductible for a given claim could reach or exceed the policy limit, in which case your insurance coverage would be depleted before all or part of any claim is paid. This argues for paying the extra premium in order that the legal costs be covered outside the policy limit.

14.1.5   When a loss occurs

Give your insurance company or companies notice of any suspected or known loss or claim in strict accordance with the terms of the policy or policies, describe the situation, and make it clear that you are giving notice. Failure to do so can leave you without coverage. Having discussed a claim situation with your agent or insurance company, even by way of written correspondence, may not constitute notice as required by the applicable policies. Give notice even if a deductible has not been met.

When you have a loss that may be covered by insurance, even if it is not certain, immediately begin planning the content of your notice to your carrier. In preparing the documentation and explanation of your potential claim, be aware that whatever information you include in your notice may be revealed in any related litigation.

Even when your insurance carrier provides lawyers for your legal defense, there may be times when you want to consult your own construction attorney for a second opinion about certain aspects of their defense.

Don't try to settle claims outside of your carrier. Your insurance company can refuse to defend you if the claim resurfaces, on the basis that your interference may have precluded the carrier's alternatives.

As soon as possible following a loss, begin to document the conditions that may have a bearing on an insurance claim, including the following:

- Time and date of the loss.
- Who was present at the time of or immediately following the loss?

- Who or what caused the loss?
- What is the nature of the loss, i.e., property damage, personal injury, etc.?
- Which policies are applicable and whom do they cover?
- Whom do you indemnify under the construction contract?
- Who indemnifies you?
- Take photographs as relevant, including time and date stamp.

You may decide to consult your construction attorney in addition to your insurance agent.

14.1.6   Insurance administration

Here are some tips about your insurance coverage and administration:

- Maintain files for your insurance policies and related notices and correspondence.
- When you have a claim, follow the exact policy requirements pertaining to giving notice to the carrier or agent, even when you know they are aware of the loss.
- Always obtain a copy of any policy that affects you and have your agent review it. Make no assumptions that you are covered. Verify.
- Require that you be named as "additional insured" on all policies by related other parties—especially subcontractors.
- Mold and other environmental coverage are limited and expensive. Discuss your exposure and your options with your agent.
- If you are required by the contract to provide certain insurance such as builder's risk and fail to do so, you may be responsible for any losses that would have been covered by the policy.
- If your application for insurance is found to contain inaccuracies after issuance of the policy, the carrier may have the right to rescind the coverage and refund your premiums. The effect will be as if the coverage never existed.
- Your insurance is only as good as your carrier. Your agent can determine a carrier's financial rating as compared with industry standards.
- Your insurance company may accept a claim, deny it, or accept it with "reservation of rights." Reservation of rights means the insurer's defense is contingent on the determination of facts yet unknown. If those facts determine that your insurance does not apply, you are left without coverage for that claim.

- Do not agree to an indemnification clause that requires you to indemnify another party against liability caused solely by that party.
- You should have someone in your office who is trained and knowledgeable in construction insurance, at least to the extent that he is aware of the areas in which he does not have all of the insurance knowledge he needs in a given situation. This may be you, your bookkeeper, or another responsible employee.
- Your field personnel should be versed in recognizing a possible claim, documenting it as described above, and notifying those in the company who have specialized knowledge.
- **Note of caution**   Be aware that you cannot recover your losses that result from your own faulty work. General liability policies have historically provided coverage for damages to the completed work of a lower-tier contractor, and/or to the work of the general contractor caused by the negligence of a lower-tier contractor. However, recent court decisions have eroded this coverage in certain states citing the faulty work exclusion. Have your agent confirm that you are protected for damages to the work of others caused by the negligence of a lower-tier contractor.

14.1.7   Certificates of insurance

Insurance certificates are issued by an insurance company to certify that it has provided the described insurance to the party named as the "Insured" on the certificate. You will be required to provide your insurance certificate to various parties at the time of contract execution and afterwards. Your agent probably has the authority to issue them upon your request. Similarly, you will require certificates from your subcontractors and other parties to a construction project.

Insurance certificates are important documents that require a certain level of management. They are evidence that you are covered under another party's insurance policies. You should adopt the following rules about insurance certificates that you require from others:

- Accept only signed original certificates, not faxes or copies, from the insurance company or its authorized agent.
- Require that the certificate name the insured—e.g., your subcontractor—exactly as he is named on your subcontract agreement (which must be his exact legal name).

- Specify that your company name be shown correctly on the certificate as the "Holder."
- Require that you be given a minimum of thirty days' written notice prior to cancellation or alteration of the insurance shown on the certificate.
- Require receipt of the proper, original certificate before your subcontractor begins field work on your project.
- State in your subcontract that you are authorized to acquire the insurance coverage required of the subcontractor at his expense if he fails to provide the coverage. However, acting on this authorization may have other implications, so consult your insurance agent or construction attorney before doing so.
- State in your subcontract that your subcontractor will indemnify and defend you for liability loss where the subcontractor is involved—even partially.
- Set up a system for easy retrieval and long-term safekeeping of insurance certificates and policies.

14.1.8 Waiver of subrogation

A waiver of subrogation rights should be included in the contracts of all parties to the construction project. This prevents one party's insurance carrier from making a claim against another party or parties on the project.

14.2 Construction surety bonds

A surety, usually an insurance company, is a company that guarantees for a fee the obligations of one party to another party. A typical case in construction is when the surety guarantees to a project owner (the "obligee") that his contractor (the "principal") will perform as required by the construction contract between the owner and the contractor. This guarantee is a "bond," made in writing. If the contractor defaults on his obligations under the construction contract, the surety is obligated to the owner to see that the contract is completed—up to the face amount of the bond. The surety will raise any incidence of non-compliance with notice or other requirements as a defense against payment under the bond. Because the bonding company is lending its credit to the principal, it will also use every legal means to recover its costs from the defaulted contractor. Courts and arbitration boards are loaded with cases arising from disagreements between parties to bonds.

There are bonds for various purposes. Performance bonds are often required by an owner as a means of ensuring that the project will be built in accordance with his plans and specifications. He may also require payment bonds, which guarantee that subcontractors and other parties to the project will be paid for the goods and services they provide. This assures the owner that these vendors will not have cause to place liens on his property. There are other bonds that are used for miscellaneous purposes in construction but most are less imposing than performance bonds and payment bonds.

Contractors may require performance and payment bonds of their subcontractors. If you ever have to make a claim against a bond for a subcontractor under contract to you, you will need to read the bond to know exactly what notice and other procedures are required of you. In your notice, describe the default situation and make it clear that you are demanding that the surety perform its obligations under the bond.

Your personal indemnification of any surety, creditor, or lender means your personal assets are at risk if you default on your obligations. Many vendors will extend credit guaranteed only by your corporation, but that is generally not the case with sureties and banks.

There is no industry standard for bonds. Both you and the owner are free to ask for terms that suit you, and the surety is free to accept or reject them. Bonds are not required by law on private construction projects except in certain states. In an owner–contractor relationship, the owner weighs his cost of the bond against the confidence he has in his contractor in deciding whether to require bonds. Once well established and able to provide evidence of financial viability to an owner's satisfaction, you may not be required to provide bonds for that owner's projects.

Chapter 15

Specializing in chain store construction

The best that an individual can do is to concentrate on what he or she can do, in the course of a burning effort to do it better.

– Elizabeth Bowen

According to a survey of 334 contractors across the United States by FMI Corporation, a management consultant to the construction industry headquartered in Raleigh, NC, contractors who differentiated themselves, i.e., specialized in a certain segment of the construction market, had a 17 percentage point better success rate in winning bids than contractors who were not differentiated: The differentiated firms had a success rate of 37 percent compared to 20 percent for the non-differentiated.

Focusing on chain store construction is an example of differentiation and there are various paths within the chain store niche. You might specialize in retail mall stores like Gap, convenience stores like 7-Eleven, the huge restaurant group that includes Starbucks and Cracker Barrel, or the "big boxes" such as Circuit City and Staples. Even banks often fit the chain store mold since competition and economics have driven them into a pattern of traditionally large "financial centers" surrounded by smaller-footprint branch banking offices clustered in neighborhoods and strip shopping areas.

As for restaurants, you only have to keep your eyes open to get an idea of the potential they hold for a contractor. Not only new eateries but also replacement and renovation projects are consistently popping up in growth areas. Chain restaurant operators, like most retailers, typically upgrade their facilities every few years in order to remain current and competitive with newer competition nearby.

I must confess that I'm partial to restaurant construction. I've built many stores in the other chain groups—convenience food stores, drug stores, branch banks, retail mall stores, and others—but restaurants were always the mainstay. It's hard to deny the potential of building restaurants when you consider that the National Restaurant Association projects the number of food service locations in the United States will increase by an average of 18,000 units per year through 2010. This number does not even include units that are continually being remodeled or rebuilt—projects that run hundreds of thousands of dollars and more and are often awarded in multiples to trusted contractors. Based on my experience, the renovation market holds at least as much potential for the contractor as the new-store segment offers.

Though smaller projects, restaurants are difficult to build (certainly, hospitals and some other projects are more complex) because of the maze of mechanical systems crammed into a tight space, the high level of finish work demanded by restaurant chains, the relatively short construction times usually allowed, and the heavy traffic that restaurant facilities must be built to endure. Not all general and subcontractors are willing or able to deal with the resulting managed chaos, so this is a road less traveled.

Other chain sub-segments or a combination of them holds similar potential. Convenience food stores/gasoline centers are ideal candidates for the chain store contractor. They are usually owned by the generally financially stable major oil companies and closely follow residential and commercial growth patterns. (Installation of the mostly underground petroleum facilities portion of these projects is often contracted separately from the building by the project owner to contractors who do this highly specialized work.)

Drug stores are typical of chain stores. There's a Walgreen's, CVS, or Eckerd's on practically every corner. Our aging population is a huge factor in this trend and can only become more so.

The list at the end of this chapter names some of the retail chain stores and restaurants operating in the United States. Major retail shopping malls often have 100–150 tenants who hire chain store contractors to build out their space after the mall is built by heavier construction contractors, who typically don't do tenant work. Continuing opportunities are created by tenant turnover and the construction of more and more shopping centers as cities and communities expand and shift to accommodate changing demographics.

While sheer size and some other aspects of chain store construction vary somewhat from one sub-segment to another, all but those requiring the largest buildings share common basic characteristics that compare favorably to the broader general construction market.

And here's more good news: The entry threshold for building chain stores in terms of contractor net worth, working capital, credit lines, and bonding capacity is manageable for most start-up contractors who have done their homework, as described in this book.

Construction contract amounts in the chain store market range from hundreds of thousands to several million dollars. In the chain store niche you can remain small as a general contractor or become larger. Some contractors I know are profitable and satisfied with a handful of half-million- to million-dollar projects a year while others find their comfort level in the $20 million to $40 million annual sales range. At the upper end of the scale, a general contractor based in Atlanta, GA, who specializes in restaurants and other chain stores across the United States does more than $200 million a year in sales. But with effective management you can make handsome profits and accumulate wealth over time even with the lower volumes of work mentioned above.

Contractors in each of these groups start from scratch. They survive for a long term only if they learn and practice sound business management such as are presented in this book.

Here are more of the advantages of specialization in chain store construction:

15.1 Improved profit potential

As a general contractor specializing in the construction of one of the chain store sub-segments, you have the opportunity to create a reliable stream of profitable, lower-risk projects, which aren't available to the same degree if you are a generalist depending on random bids you learn about through project information services such as McGraw-Hill Companies, Inc.'s, "Dodge Reports," and on architects you've worked with.

Somewhat surprisingly, the chain store construction niche is not overcrowded. During my years focusing on narrow chain store segments, new construction firms appeared from time to time and others dropped out, but when I sold my company the level of competition was not much different than it had been twenty-five years earlier when I was getting started. Much of my firm's work was negotiated with repeat customers, but when bidding was required there were rarely more than three bidders, and often only two of us. I was usually familiar with the other bidders, even in a market that included a dozen or so states. Having some insight into your competitors' bidding practices is an important advantage when you're finalizing a bid.

15.2   Continuing relationships

Specialization offers you unusual potential to develop continuing relationships with companies that generate numerous projects year after year.

Project owners and their construction representatives are no different than the rest of us in terms of human nature: Other factors being equal, they favor contractors who make their jobs run smoothly and whose project managers and superintendents they find likeable on a personal level. A great personality alone may not land you the business, but being unpleasant to deal with can spoil your plans all by itself.

If you hire the right people, train them, provide them with the tools and support they need to do their jobs, and instill the values that serve both your company and your customers well, as discussed earlier, the advantages of contracting with your firm will be apparent to project owners.

In contrast, lowest-bid one-of-a-kind projects in the private and public sectors alike often are a source of adversarial relationships. Projects awarded on price alone yield low profit margins, on average, compared to the opportunities available in the restaurant and other chain store construction business.

15.3   Chain operators favor niche contractors

The expanding multi-unit chain operator prefers the general contractor who specializes in his field. This is because he knows that this contractor:

- Is accustomed to the tight scheduling typically required
- Produces quality work in optimum time
- Knows and uses subcontractors who are familiar with this kind of work
- Builds his business around repeat customers and goes the extra mile to earn it
- Speaks the industry lingo
- Requires less handholding
- Can build multiple projects at once
- May work beyond the chain's local area in adjacent or more distant states, a convenience to a fast-moving chain

These factors combine to reduce the chain owner's or franchisee's risk, and his or his construction representative's time and expense. And the contractor benefits as well: Quite often the chain owner doles out and negotiates his projects in a given region among two or three general contractors he's learned he can depend on. This presents the contractor who is willing to tune in to and accommodate his customers' special wants and needs—i.e., his pet likes and dislikes for his projects that are not shown in the drawings and specifications (they all have them)—the opportunity to cultivate a loyal customer who will not as likely be tempted by other contractors playing the price card to get their foot in the door.

15.4   Fewer parties in the mix

Chains usually have in-house architectural and engineering departments that design their own buildings and generate their own reproducible construction drawings. Thus their use of outside architects and engineers is usually limited to conforming these "stock" plans to local codes and specific site conditions.

The contractor may become involved in the project only after the local professionals have completed their work, and therefore may have no direct contact with them unless questions or problems related to their work arise during bidding or construction. So you deal directly with the owner or his construction representative on practically all matters related to the project.

There are important differences between owner–architect–contractor jobs and owner–contractor jobs. In the first case, the architect is a middleman the contractor works with instead of directly with the owner or owner representative, and this arrangement is less than ideal for you as the contractor. The architect often must get time-consuming approval from the owner before answering your questions. According to the provisions of some widely used owner–contractor agreement forms, you may be required to rely on the architect's instructions but the owner may not be bound by them. You can imagine problems this may cause, for example, getting paid for a change required by the architect that the owner disputes for some reason. The architect is expected to be neutral, but his role may put him in an awkward position at times because he is paid by the project owner.

I much prefer working directly with the owner, a process that is more streamlined and holds less potential for miscommunication. However, this is in no way an indictment of architects. The construction industry would not exist without them. Almost without exception, those I've known and worked with have been competent, fair, and interested in keeping their projects moving.

15.5   Reliable cost database

Although the multiple building designs within any given chain vary somewhat, in most cases a consistent theme is present throughout the details. Whether you're building the same building over and over or some of its variations within a chain, you have the opportunity to develop a cost database for that chain of stores and refine it through repetition. This database becomes invaluable in bidding future projects. For instance, it allows you to trim your bid to the barest minimum with confidence when necessary, making it tough for a newcomer contractor to compete with you.

15.6   Reduced risk

The consequences of construction problems run the gamut from merely difficult to unfinished projects and even bankruptcy of one or more of the parties. These risks to the contractor are generally not as great in chain store construction, partly due to shorter job duration.

15.7   So why are so many contractors missing out on the chain store niche?

It may be that chain store construction doesn't seem "big-time" enough for many contractors; after all, even the most exciting chain store project is not likely to make the cover of *Engineering News Record*. I have seen contractors jump into restaurants and quickly pop back out—probably wondering how we in the business meet the demands. Unfortunately for those contractors, they were unable or unwilling to conform themselves to this kind of construction.

15.8   A note about the construction industry as a whole

In the United States, much has been written and debated about the loss of jobs and even significant portions of some industries to certain foreign countries in which labor is cheaper. This should not be a concern within the construction industry; the construction of buildings and other structures will continue to require plumbers, electricians, grading contractors, and the entire spectrum of building trades to perform work *on-site*—meaning construction firm owners, managers, superintendents, and other key personnel will always be needed at home.

Construction is a huge part of US economy and is made up of mostly small firms. In *Quick Facts About the Construction Industry*, Chief Economist Kenneth Simonson of the AGC reports that the average number of employees for all US construction firms in 2002 was only nine per firm, and that 91 percent of all construction firms had fewer than twenty employees.

In a paper written for The Brookings Institution, Arthur C. Nelson says that in the year 2030, "half of the buildings in which Americans live, work, and shop will have been built after 2000," and that more than 60 percent of the commercial and industrial space in 2030 will be less than thirty years old. It is hard to imagine an industry offering more promise than construction.

Industry prognosticators fear that too few qualified candidates are training for and entering the field to supply the demand in skilled, managerial, and leadership positions. This means young people in the United States can benefit from continuing exciting opportunities in construction that are limited only by their own imaginations.

While the numbers vary from country to country, the US construction industry is not unique in making a healthy contribution to its nation's economy. The construction industry is among the top contributors to the Gross Domestic Product (GDP) and to employment rolls in the United Kingdom, the European Union, and Canada as well as the United States, as shown in the cross-reference chart in Appendix 3.

Partial list of retail chain stores			
7-Eleven	Crate & Barrel	Lenscrafters	Safeway
AAMCO	CVS	Limited	Sam Goody
Ace Hardware	Dollar General	Linens 'n Things	Sam's Club
Advance Auto Parts	Dollar Tree	Lowe's	Service Merchandise
Albertsons	Eckerd Drugs	Luxottica	Sharper Image
Allstate	Eddie Bauer	MasterCuts	Shell Oil
Ann Taylor	Exxon	Medicine Shoppe	Sherwin Williams
AutoZone	EZ Mart	Meineke	Speedway
Banana Republic	Family Dollar	Men's Warehouse	Sports Authority
Barnes & Noble	Famous Footwear	Michael's	Staples
Bass Pro Shops	Fashion Bug	Montessori	State Farm
Bed Bath & Beyond	Firestone	Neiman Marcus	Stein Mart
Berkshire-Hathaway	Foot Locker	Nordstrom	Sunglass Hut
Best Buy	Gap	Office Depot	Super Value
Big Lots	GNC	OfficeMax	SuperCuts
Blockbuster	Goodyear	Old Navy	Suzuki
Bloomingdale's	Hallmark	Payless ShoeSource	T. J. Maxx
Bombay Company	Hampton Inn	Pep Boys	Target
Borders	Hickory Farms	Petco	Tom Thumb
BP	Home Depot	PetSmart	Toys R Us
Car Dealerships	John Deere	Pier 1 Import	True Value
Century 21	Kinko's	Publix	Verizon
Circle K	Kmart	QuikTrip	Waldenbooks
Circuit City	Kohl's	Radio Shack	Walgreen
CompUSA	Kroger	ReMax	Wal-Mart
Costco	La Petite	Rite Aid	Williams-Sonoma
Cracker Barrel	Lane Bryant	Ritz Camera	Zales

Partial list of chain restaurants

A & W	Chart House	Golden Corral	Panera Bread
Applebee's	Checkers	Green Burrito	Papa John's Pizza
Arby's	Cheesecake Factory	Grill Concepts	Papadeaux Seafood
Atlanta Bread Co	Chick-Fil-A	Haagen-Dazs	Pizza Inn
Bahama Breeze	Chili's	Hardee's	Popeyes
Baskin-Robbins	Chipotle Mexican Grill	Houlihan's	Quizno's
Ben & Jerry's	Chuck-E-Cheese	Huddle House	Red Lobster
Bertolini's	Church's Chicken	IHOP	Romano's Macaroni Grill
Biaggi's	Cinnabon	Jack In The Box	Ruby Tuesday
Big Boy	Corner Bakery Café	Johnny Rockets	Ryan's Family Steak
Blimpie	Cracker Barrel	KFC	Sbarro
Bob Evans Farms	Dairy Queen	Krispy Kreme	Seattle's Best Coffee
Bojangle's	Del Taco	Krystal	Sonic Drive In
Boston Market	Denny's	Little Caesars	Spaghetti Warehouse
Brio Tuscan Grille	Domino's Pizza	Logan's Roadhouse	Starbucks
Bruegger's	Donato's	Long John Silver's	Subway
Burger King	Dunkin' Donuts	Longhorn Steakhouse	Taco Bell
Cajun Café	Einstein's	Maggiano's Little Italy	TCBY
Cajun Kitchen	El Pollo Loco	McDonald's	T. J. Cinnamon's
Cal. Pizza Kitchen	Fazioli's	Moe's Southwest Grill	Waffle House
Captain D's	Friendly's Ice Cream	Olive Garden	Wendy's
Carl's Jr	Fuddrucker's	On The Border	Whataburger
Carrabba's	Godfather Pizza	Outback Steakhouse	White Castle

Appendix 1

If you're just getting started . . .

. . . a good beginning is half the business.

– Plato

This chapter draws from various parts of the book and may be used as a quick reference for the start-up contractor.

- Develop a business plan that aggressively but accurately reflects your firm and its potential. You probably have already done this.
- With the help of an outside certified public accountant (CPA), establish billing and payment policies and procedures, decide on accounting software, create a chart of accounts, and set up filing and retrieval systems.
- Meet with a business lawyer who, along with your CPA, can help you decide on your business structure, most likely an "S" corporation or a "C" corporation.
- Implement a computerized construction-accounting system.
- If in the beginning you are limited to hiring only one person to handle reception, administrative, and bookkeeping functions, lean toward a person with bookkeeping experience.
- Obtain an employee identification number (EIN) from the IRS at www.irs.gov.
- Know how, when, and where to remit your employees' payroll withholdings to the federal and state governments. Remit these payments on time to avoid penalty.
- When possible, use forms published by AGC, ABC, or AIA and modify them as required. Have a construction attorney review contracts before you sign them unless you know what to look for in contract terms and provisions.
- Register your corporation with your state's secretary of state and other agencies as required. This applies to all states in which you will do business.

- Get a business license.
- Get a contractor professional license if required where you will do business. This may take several weeks and may be required before you submit a bid on a project.
- Establish vendor accounts for building materials. Pay bills when due in order to establish good credit history.
- Decide who is responsible and accountable for approving incoming invoices from subcontractors and other vendors, and what accompanying documentation is required.
- Determine in advance each customer's invoicing requirements to avoid delay in applying for and receiving payment.
- Establish banking relationships, including a line of credit if possible. Use the SBA for financial assistance if needed.
- Sign all checks yourself.
- Obtain business and project-specific insurance coverage through a construction insurance agent.
- Establish field operating policies and procedures and name who has the responsibility and authority for carrying them out.
- Track general expenses (payroll, rent, car expense, insurance, etc.) bank balances, job costs, contract amounts, income, payables, and receivables from the first day forward.
- Maintain separation between your personal and company business.
- Determine additional requirements as required by your particular circumstances.
- Read this book for more information.

YOUR CONSTRUCTION COMPANY
VITAL SIGNS REPORT SPREADSHEET
2008 by week

A	B	C	D	E	F
Week Ending	Cash and Money Market Accounts	Accounts Receivable	Job Costs Payable	Other Payables (General Overhead, Loan Pmts, etc.)	VITAL SIGNS INDEX B+C–D–E
	Input Value ($)	Input Value ($)	Input Value ($)	Input Value ($)	Calculated Value ($)
4-Jan-08	678,226	1,632,307	827,816	19,615	1,463,102
11-Jan-08	645,342	1,539,546	810,027	18,845	1,356,016
18-Jan-08	592,231	1,531,899	811,353	19,566	1,293,211
25-Jan-08	880,127	1,549,815	779,981	27,023	1,622,938
1-Feb-08	801,081	1,408,960	759,610	19,931	1,430,499
8-Feb-08	726,775	1,403,961	788,383	18,254	1,324,099
15-Feb-08	706,710	1,283,917	765,809	18,686	1,206,132
22-Feb-08	853,680	1,464,501	672,168	26,525	1,619,488
29-Feb-08	761,703	1,521,350	869,167	19,797	1,394,089
7-Mar-08	727,718	1,506,276	1,049,829	18,163	1,166,002
14-Mar-08	1,159,736	1,315,510	1,092,479	19,462	1,363,305
21-Mar-08	895,603	1,815,915	946,249	18,541	1,746,728
28-Mar-08	1,134,971	1,722,860	1,214,753	27,409	1,615,669
4-Apr-08	934,959	1,742,776	1,518,889	18,717	1,140,129
11-Apr-08	730,345	1,906,022	1,483,258	19,191	1,133,919
18-Apr-08	159,703	2,159,249	1,130,678	18,829	1,169,446
25-Apr-08	762,233	2,480,883	1,361,674	27,288	1,854,155
2-May-08	813,815	2,320,671	1,582,845	17,844	1,533,797
9-May-08	1,023,985	2,000,889	1,453,778	18,976	1,552,121
16-May-08	1,068,079	1,925,435	1,716,397	18,311	1,258,807
23-May-08	897,267	2,246,721	1,179,701	26,977	1,937,309
30-May-08	753,471	2,421,431	1,368,858	19,502	1,786,542
6-Jun-08	1,181,952	1,980,728	1,468,283	18,614	1,675,783
13-Jun-08	1,264,124	1,983,964	1,546,214	18,199	1,683,676
20-Jun-08	843,419	2,450,558	1,303,960	17,878	1,972,139
27-Jun-08	694,306	2,845,820	1,589,662	27,853	1,922,610
4-Jul-08	813,291	2,783,593	1,904,000	19,228	1,673,657
11-Jul-08					–
18-Jul-08					–
25-Jul-08					–
1-Aug-08					–
8-Aug-08					–
15-Aug-08					–
22-Aug-08					–
29-Aug-08					–
5-Sep-08					–
12-Sep-08					–
19-Sep-08					–
26-Sep-08					–
3-Oct-08					–
10-Oct-08					–
17-Oct-08					–
24-Oct-08					–
31-Oct-08					–
7-Nov-08					–
14-Nov-08					–
21-Nov-08					–
28-Nov-08					–
5-Dec-08					–
12-Dec-08					–
19-Dec-08					–
26-Dec-08					–

JOB PRE-STARTUP CHECKLIST

Job Name_____ Job Nbr._____ PM_____ Supt_____

Item ("Owner" refers to the project owner.)	Required? Y/N	PM Sign Off When Item Satisfied	Date Item Satisfied	Notes
Legal & Administrative				"Owner" refers to the project owner.
Owner–Contractor agreement signed				
Owner financing verified				
Written Notice to Proceed from the Owner				
Cost codes/schedule of values to bookkeeping				
Construction schedule to all parties				
Pre-construction lien rights notice to Owner				
All contracts distributed in-house and to others				
Owner's billing requirements, forms				
Supplier/vendor purchase orders issued				
Insurance				
Business insurance in place (CGL, WC, etc.)				
Job-specific coverages in place				
If Builders Risk by Owner, is policy received?				
If Builders Risk by GC, is policy received?				
Subrogation waiver by all parties				
Plans & Specifications				
Contract dwgs/revision nbrs. verified				
Licenses & Permits				
Secretary of State registration				
State/local general contractor license				
Business license				
Building permit/inspection card in hand				
Permit dwgs received				
Demolition permit				
Environmental permit				
Other permits: _____				
Easements				
Municipal bonds				
Subcontractors				
Contact info for all subs and vendors				
Subcontracts signed by both parties, delivered				
Billing requirements to subs				
Subs' original Cert. of Insurance per spec				
Subcontractor bonds				
Pre-construction meeting with subs and vendors				
Temporary Facilities and Utilities				
Office trailer				
Storage trailer				
Dumpster				
Toilet				
Electric (temp. power pole installed/inspected)				
Telephone				
Broadband internet service				
Water				
Undrgrnd util. located (Call Before You Dig)				
Other_____				
Sitework				
Soils report				
Erosion control				
Property corner pins placed by Owner				
Corner pins verified by Contractor				
Elevation benchmark established by Owner				
Benchmark verified by Contractor				
Site/Grading plan verified to existing conditions				
Testing firm hired				
Local Accounts Established				
Building materials/lumber				
Concrete				
Hardware				
Tool rental				
Temporary labor				

Note: This list is not comprehensive. Modify as required by project conditions.

YOUR CONSTRUCTION COMPANY
Project Closeout Checklist

PROJECT _____

PM _____   SUPT _____

	Date / Initials
All work complete	
All punch lists signed off by Contractor and Owner, dated, designated as "Final"	
As-built drawings	
Warranty documentation to/from all parties	
Manufacturer's owner manuals and documentation to owner	
Inspection placards	
Certificate of occupancy	
All systems operated and tested	
Demobilization: field office; storage; signs; mailbox; forwarding address; temporary utilities; rented items; company equipment	
Final lien waivers/releases from vendors, subcontractors, sub-subcontractors	
Final claims waivers/releases from vendors, subcontractors, sub-subcontractors	
Final billing to owner	
Final payments to vendors	
Jobsite records delivered to corporate office including logs, test reports, field change orders, inspection placards, as-built drawings, lien waivers and releases, punchlists, subcontractor back charges	
Discontinuance of insurance coverages specific to project as appropriate	
Archive of retained documents organized by destroy dates	
Final accounting and reports to management	
Outstanding claims	
Final clean-up	

YOUR CONSTRUCTION COMPANY
Chart of General & Administrative Expenses Accounts (Basic)

Accounting—outside CPA
Advertising and marketing
Advertising—help-wanted
Automobile/truck fuel
Automobile/truck lease
Automobile/truck maint and repair
Bank account expense
Bidding expense not incl in salaries and wages
Business meals—company business purposes
Computer software
Depreciation
Entertainment—customers and prospects
Insurance—autos and trucks
Insurance—commercial general liability
Insurance—employee fidelity
Insurance—employer liability
Insurance—worker's comp
Insurance—other
Interest on loan 1
Interest on loan 2
Internet service—monthly expense
Legal expense—corporate
Legal expense—registered agent
Legal expense—project specific (unless chg'd to project)
Licenses and regulatory—business license
Licenses and regulatory—professional contractor license
Licenses and regulatory—secretary of state registration
Licenses and regulatory—other
Miscellaneous expense (minor)
Office maint and repair
Office machine/computer lease/rental
Office machines—maint and repair
Office rent
Office supplies
Office utilities
Postage and overnight mail service
Professional consultant fees and expenses
Recreational functions, Christmas parties, etc.
Salaries and wages—permanent employees (excl job cost)
Salaries and wages—temporary help
Taxes—federal and state income taxes
Taxes—payroll
Taxes—other
Telephone basic monthly charges
Telephone long distance
Telephone—cellular
Training and education—professional and trades
Training and education—safety
Travel—air fare
Travel—car rental
Travel—lodging
Travel—meals

Letter of Introduction

YOUR CONSTRUCTION COMPANY
123 Your Street
Dallas, TX

Tel. 012-345-6788
Fax 012-345-6789

Email myname@_____.com
Web site www._____.com

July 18, 2008

Mr. James Smith
Director of Construction
Growing Chain Store, Inc.
Main Street
Chicago, Ill

Re: Your Magnolia and Pine Project, Atlanta, GA

Dear Mr. Smith:

As we discussed on the phone this morning, I will appreciate the opportunity to submit a bid on your planned new store at Magnolia and Pine Streets in Atlanta. As your contractor, my goal will be your complete satisfaction with every aspect of our performance.

As a way of introduction, I've enclosed the following information:

- A brochure that briefly describes our company
- Our Contractor Qualification Statement
- A list of several of our completed as well as under-construction projects

As indicated in the enclosures, we specialize in projects similar to yours. Additional information about my company is available on our Web site, www._____.com.

Thank you for your consideration. I will follow up with you in a few days.
Sincerely,

John Davis
President

Appendix 2

Useful Web site links

Dun & Bradstreet business statistics database	www.dnb.com
American Franchise Association	www.franchisee.org
American Institute of Architects	www.aia.org
American Institute of Architects London/UK	www.aiauk.org
American Subcontractors Association	asaonline.com/Web/index.aspx
Americans with Disabilities Act	www.usdoj.gov/crt/ada/adahom1.htm
Architects Council of Europe	www.ace-cae.org
Associated Builders and Contractors	www.abc.org
Associated General Contractors of America	www.agc.org
Blue Book of Building and Construction	www.thebluebook.com
Building Science Corporation	www.buildingscience.com
Bizfilings business registration service	www.bizfilings.com
Business Owner's Toolkit	www.toolkit.cch.com
Canadian Construction Association	www.cca-acc.com
Canadian Federation of Independent Business	www.cfib.ca
CEOExpress current business information	www.ceoexpress.com
Corporate Executive Board leadership resourse	www.executiveboard.com
Corporate Leadership Council leadership resourse	www.corporateleadershipcouncil.com
Dun & Bradstreet business statistics database	dbml2.zapdata.com
European Construction Federation	www.fiec.org
European Council for Small Business and Entrepreneurship	www.ecsb.org
International Franchise Association	www.franchise.org
Jobpower construction software	www.jobpow.com
Lebhar-Friedman retail store newspapers	www.lf.com

Nation's Restaurant News newspaper	www.nrn.com
National Federation of Independent Business (US)	www.nfib.com
National Restaurant Association	www.nationalrestaurant association.com
North American Industry Classification System (NAI) CS	www.naics.com
QSR *online* foodservice magazine	www.qsrmagazine.com
Retail Construction Magazine	retailconstructionmag.com
Royal Architectural Institute of Canada	www.raic.org
Royal Institute of British Architects	www.riba.org
SCORE non-profit business operations resource	www.score.org
State Law Matrix codes and information	www.agc.org
Stores retail industry magazine	www.stores.org
UK Construction Confederation	www.thecc.org.uk
UK Small Business Service	www.sbs.gov.uk/sbsgov
Uniform Commercial Code (US)	www.law.cornell.edu/ucc
US Bureau of Economic Analysis	www.bea.gov
US Census Bureau	www.census.gov
US Department of Labor	www.dol.gov
US Environmental Protection Agency	www.epa.gov
US Government printing office	www.gpoaccess.gov
US Government Web sites portal	www.firstgov.gov
US Internal Revenue Service	www.irs.gov
US Occupational Safety and Health Administration (OSHA)	www.osha.gov
US SBA small business advocate	www.sba.gov/advo
US Small Business Administration	www.sba.gov

Appendix 3

Regional cross reference of construction-related organizations[1]

	US	UK	Canada	Europe
Major Construction and Contractor Organizations/Industry Advocates[2]	Associated General Contractors of America (AGC) www.agc.org Associated Builders and Contractors (ABC) www.abc.org	The Construction Confederation www.thecc.org.uk	Canadian Construction Association (CCA) www.cca-acc.com	The European Construction Federation (FIEC). Each member country's contact information is available at the FIEC Web site (www.fiec.org)
Description of Construction Organizations	AGC and ABC represent non-residential member contractors through their respective chapter associations located around the US	The Construction Confederation is an umbrella group for construction-industry organizations within the UK, who in turn directly represent member companies and related groups	Composed of non-residential contractor associations. Contractors and related professionals and sectors may join through member local and provincial construction associations	FIEC is an umbrella group for construction industry associations in 27 countries, who in turn represent member companies in their respective countries
Member Countries	US	UK	Canada	Austria, Belgium, Bulgaria, Croatia, Cyprus, Czech Republic, Denmark, Estonia, Finland, France, Germany, Great Britain, Greece, Hungary, Ireland, Italy, Luxembourg, Netherlands, Norway, Portugal, Romania, Slovakia, Slovenia, Spain, Sweden, Switzerland, Turkey. Source: FIEC www.fiec.org

	US	**UK**	**Canada**	**Europe**
Construction Industry Contribution to GDP	8% in 2004 Source: AGC www.agc.org	10% in 2003 Source: The Construction Confederation www.thecc.org.uk	11% In 2004 Source: National Research Council Canada www.nrc-cnrc.gc.ca	10% in 2004 Source: FIEC www.fiec.org
Percent of Work Force in Construction	5% in 2004 Source: AGC www.agc.org	7% in 2003 Source: Construction News www.cnplus.co.uk	6% (projected) in 2005 Source: CCA www.cca-acc.com	7% in 2004 Source: FIEC www.fiec.org
Small Business Assistance/ Advisory Groups/ Organizations [3]	US Small Business Administration (SBA) www.sba. gov National Federation of Independent Business (NFIB) www.nfib.com	Small Business Service (SBS) www.sbs.gov.uk	Canadian Federation of Independent Business (CFIB) www.cfib.ca	European Council for Small Business and Entrepreneurship (ECSB) www.ecsb.org
Representative Architect /Design Organizations	American Institute of Architects (AIA) www.aia.org	Royal Institute of British Architects (RIBA) www.riba. org AIA London/UK www.aiauk.org	Royal Architectural Institute of Canada (RAIC) www.raic.org	Architects Council of Europe (CAE) www.ace-cae.org

Notes:
1. The information contained herein is believed to be correct but cannot be guaranteed.
2. The construction industry engenders many peripheral groups. Only a few are shown here.
3. Not an exhaustive list of small business assistance organizations.

Appendix 4

Potential questions for interviewing job applicants[1]

Questions to reveal integrity/honesty/trustworthiness

- Discuss a time when your integrity was challenged. How did you handle it?
- What would you do if someone asked you to do something unethical?
- Have you ever experienced a loss for doing what is right?
- Have you ever asked for forgiveness for doing something wrong?
- In what business situations do you feel honesty would be inappropriate?
- If you see a co-worker doing something dishonest, would you tell your boss? What would you do about it?

Questions to reveal personality/temperament/ability to work with others

- If you took out a full-page ad in a newspaper and had to describe yourself in only three words, what would those words be?
- How would you describe your personality?
- What motivates you the most?
- If I call your references, what will they say about you?
- Do you consider yourself a risk taker? Describe a situation in which you had to take a risk.
- What kind of environment would you like to work in?
- What kinds of people would you rather not work with?
- What kinds of responsibilities would you like to avoid in your next job?
- What are two or three examples of tasks that you do not particularly enjoy doing? Indicate how you remain motivated to complete those tasks.

- What kinds of people bug you?
- Tell me about a work situation that irritated you.
- Have you ever had to resolve a conflict with a co-worker or client? How did you resolve it?
- Describe the appropriate relationship between a supervisor and the subordinates.
- What sort of relationships do you have with your associates, both at the same level and above and below you?
- How have you worked as member of teams in the past?
- Tell me about some of the groups you've had to get cooperation from. What did you do?
- What is you management style? How do you think your subordinates perceive you?
- As a manager, have you ever had to fire anyone? If so, what were the circumstances and how did you handle it?
- Have you ever been in a situation where a project was returned for errors? What effect did this have on you?
- What previous job was the most satisfying and why?
- What job was the most frustrating and why?
- Tell me about the best boss you ever had. Now tell me about the worst boss. What made it tough to work for him or her?
- What do you think you owe to your employer?
- What does your employer owe to you?

Questions to reveal past mistakes

- Tell me about an objective in your last job that you failed to meet and why.
- When is the last time you were criticized? How did you deal with it?
- What have you learned from your mistakes?
- Tell me about a situation where you "blew it." How did you resolve or correct it to save face?
- Tell me about a situation where you abruptly had to change what you were doing.
- If you could change one (managerial) decision you made during the past two years, what would that be?

- Tell me of a time when you had to work on a project that didn't work out the way it should have. What did you do?
- If you had the opportunity to change anything in your career, what would you have done differently?

Questions to reveal creativity/creative thinking/problem solving

- When was the last time you "broke the rules" (thought outside the box) and how did you do it?
- What have you done that was innovative?
- What was the wildest idea you had in the past year? What did you do about it?
- Give me an example of when someone brought you a new idea, particularly one that was odd or unusual. What did you do?
- If you could do anything in the world, what would you do?
- Describe a situation in which you had a difficult (management) problem. How did you solve it?
- What is the most difficult decision you've had to make? How did you arrive at your decision?
- Describe some situations in which you worked under pressure or met deadlines.
- Were you ever in a situation in which you had to meet two different deadlines given to you by two different people and you couldn't do both? What did you do?
- What type of approach to solving work problems seems to work best for you? Give me an example of when you solved a tough problem.
- When taking on a new task, do you like to have a great deal of feedback and responsibility at the outset, or do you like to try your own approach?
- You're on the phone with another department resolving a problem. The intercom pages you for a customer on hold. Your manager returns your monthly report with red pen markings and demands corrections within the hour. What do you do?
- Describe a sales presentation when you had the right product/service, and the customer wanted it but wouldn't buy it. What did you do next?

Miscellaneous good questions

- How do you measure your own success?
- What is the most interesting thing you've done in the past three years?
- What are your short-term or long-term career goals?
- Why should we hire you?
- What responsibilities do you want, and what kinds of results do you expect to achieve in your next job?
- What do you think it takes to be successful in a company like ours?
- How did the best manager you ever had motivate you to perform well? Why did that method work?
- What is the best thing a previous employer did that you wish everyone did?
- What are you most proud of?
- What is important to you in a job?
- What do you expect to find in our company that you don't have now?
- Is there anything you wanted me to know about you that we haven't discussed?
- Do you have any questions for me?

Continued on next page

These materials are provided "as is," without warranty of any kind, express or implied, including but not limited to: Warranties of performance, merchantability, fitness for a particular purpose, accuracy, omissions, completeness, currentness, and delays. The user's exclusive remedy, and the entire liability of Ceridian Corporation, its affiliates and/or contributors under this agreement, if any, for any claim(s) for damages relating to use of this publication is limited to the price paid for the publication which is the basis of the claim(s).

In no event shall Ceridian Corporation, its affiliates and/or contributors be liable to the user for any claim(s) relating in any way to
(i) any decision made or action taken by the user in reliance on any material or representation in the publication or
(ii) any lost profits or other consequential, exemplary, incidental, indirect, or special damages relating in whole or in part to the use of the publication, even if Ceridian Corporation, its affiliates, and/or contributors have been advised of the possibility of such damages.

Glossary

401(k) A retirement plan that is authorized in section 401(k) of the US Internal Revenue Code, which allows an employee to have part of his income regularly withheld and invested with taxes deferred until the money is withdrawn.

ABC See Associated Builders and Contractors, Inc.

ADR See alternative dispute resolution.

AGC See Associated General Contractors of America.

AIA American Institute of Architects.

Alternative dispute resolution A method used to settle disagreements between parties in order to avoid litigation; may or may not be binding on the parties.

Application for payment A party's request for payment for services performed to date. Commonly called a "draw."

As-built drawings Drawings made by the contractor and subcontractors showing the location of underground utilities and other details not shown on the construction drawings.

Associated Builders and Contractors, Inc. (ABC) Construction industry organization and advocate.

Associated General Contractors of America (AGC) Construction industry organization and advocate.

Bid documents The drawings, specifications, reports, instructions, and other materials provided by an owner to a contractor on which the contractor bases his bid for a given project.

Bid notes Clarifications and other information included in a bid proposal by a contractor, when permitted by the owner.

Bond An agreement by which one party, a surety, guarantees to a second party that a third party will meet its contractual obligations.

Brand The marketing name a chain uses for one or a group of its stores.

Certificate of occupancy Authorization by a city or other governing agency certifying that a building may be occupied for its intended purpose.

Chain store   One of a series of stores owned or branded by the same company or its franchisees or affiliates.

Change order   Procedure or document by which a contract may be modified.

Chart of accounts   A chart explaining the numerical codes that identify the ledger accounts in an accounting system.

COBRA   The Consolidated Omnibus Budget Reconciliation Act of 1985. Applies to employers with 20 or more employees. Requires employers to offer certain individuals who would otherwise lose benefit protection the option of continuing to have group health care plan coverage.

Construction representative (owner's representative)   A project owner's employee or other agent who contracts for and oversees the owner's construction projects in a given area. Small franchisees may not have a construction representative, relying instead on the franchisor who provides these services for a fee that is included in or as an addition to the basic franchise fee.

Contractor   The general contractor; a construction firm; the owner of the construction firm or his representatives.

Corporation   Legal entity.

D&B   See Dun & Bradstreet.

Database   A systematically arranged collection of computerized cost or other data, structured for automated retrieval or manipulation into different formats.

Differentiation   Focus by a contractor on a certain construction market niche; specialization.

Dodge Report   A service by publisher McGraw-Hill that provides information about bidding opportunities, bid results, and other current information useful to contractors and others in related fields.

Draw   See application for payment.

Dun & Bradstreet   A national source of detailed information on businesses, including industry represented, sales and financial data, number of employees, etc.

Easement   A right granted by a property owner that gives another party limited use of the owner's property.

Entrepreneur   A person who starts up a new for-profit business.

Force majeure   An unexpected or uncontrollable event that may excuse an affected party's failure to perform as required. As related to a construction agreement, examples of events that might qualify as *force majeure* are acts of war that result in the unavailability of construction materials, earthquake, or hurricane.

Franchise   A right to sell a company's products and/or use its name in specified geographic area.

Franchisee   An independent business owner who has acquired the right to sell a franchisor's products and/or use the franchisor's name in his own business.

Franchisor (franchiser)   A person or organization that sells to others (its franchisees) the right to use its name and sell its products.

G&A   See general and administrative expense.

GDP   See gross domestic product.

General and administrative expense   Usually called G&A or overhead, this is the portion of a construction firm's business expenses that cannot be easily allocated to a certain job or jobs, such as office rent, utilities, and administrative salaries.

General conditions expense   Construction cost that cannot be allocated to any specific part of the project, but rather applies generally to all aspects of the project; also known as project overhead.

General conditions   A part of a construction agreement that describes the rights and responsibilities of the parties.

General contractor   A party entering into an agreement with a project owner to take overall responsibility for the completion of a project in accordance with certain plans and specifications.

Good faith   Honesty of intention.

Gross domestic product   The total value of all goods and services produced in a country in a year, offset by net income from investment in foreign countries.

Indemnification agreement   An agreement by which one party assumes the obligations of another party.

Individual Retirement Account (IRA)   A US, tax-deferred retirement account available to individuals.

IRA   See Individual Retirement Account.

Lien waiver   A legal document by which a party relinquishes any rights he may have to place a lien on another party's property.

Lien   See materialman's/mechanic's lien.

Line item   The element of cost included on one line in a bid, cost report, or schedule of values.

Litigation   Pursuit of a lawsuit.

Materialman's/mechanic's lien A document and procedure by which a party gives legal notice that he is due payment for materials or labor he has provided to the property described in the document.

Municipal bond A bond issued by a state, city, or local government to raise capital.

NAICS See North American Industry Classification System.

Negotiated contract Construction agreement awarded to a contractor without competitive bidding.

Niche An area of the market specializing in a particular type of product. In construction, it refers to any subdivision of the construction market.

North American Industry Classification System (NAICS) A numbering system used by government agencies to identify companies by industry and sub-industry. Many government reports are categorized by NAICS code number.

Overhead See general and administrative expense.

Owner Refers to the project owner unless otherwise designated. See "Project owner."

Party A person, firm, or group of people acting together as one of the principals in an agreement. A construction agreement usually consists of two principal parties: the "owner" and the "contractor."

Payables Amounts to be paid.

Penalty clause A contract provision that obligates the contractor to forfeit a stipulated portion of the contract price to the owner if the contractor does not meet schedules specified in the contract.

PEO Professional Employer Organization.

Project manager A project manager is the contractor employee to whom the contractor assigns authority and responsibility for accomplishing the contractor's duties under a construction agreement, usually through jobsite superintendents. Project managers in small construction firms may also prepare bids for new work and perform other functions as well.

Project overhead See general conditions expense.

Project Construction work as proposed by an owner or under construction by a contractor.

Project owner A franchisor, franchisee, or independent business person who contracts with contractors to build a facility. The person or company the contractor looks to for payment for his services, i.e., the contractor's customer. The project owner may or may not be the same legal entity as the property owner. Referred to herein as "owner" or "project owner."

Property Owner The owner of the property on which a construction project is built; may or may not be the project owner.

Prototype A building planned by an owner on which future buildings will be modeled with or without variations.

Punch list A formal document or a contractor's internal note identifying incomplete or faulty work that must be completed or corrected.

Quantity takeoff The list and amount of materials and labor required to complete a project, used to prepare the contractor's bid.

Receivables Amounts due from others.

Retainage A portion of an earned amount withheld by one party from another party until certain work is completed in accordance with the contract documents.

SBA See Small Business Administration.

Schedule of values A document that specifies the allocation of the total contract amount to its various parts, used by contractor and owner as a basis for contractor's applications for payment.

Small Business Administration (SBA) A US Government agency intended to provide business advice and government-backed financial assistance and to act as an advocate in Washington for small business.

Small business Building contractors whose gross annual sales don't exceed $28,500,000, which makes them eligible for SBA programs.

Soils report A formal report that describes the technical characteristics of the soils present on a specified site, prepared by a soils engineer and used by owners and contractors to evaluate what work may be required below the surface of the site. A soils report should be included in the bid documents.

Specialty contractor A contractor specializing in a specific trade such as plumbing or electrical.

Start-up A company that is in its early stages of operation.

Store Sometimes used to refer to a chain store.

Subcontractor A party entering into an agreement with a general contractor to complete a part of the general contractor's agreement with the owner.

Subrogation Subrogation occurs when an insurance company sues a third party after it has paid off its injured claimant who has a right of claim against that third party.

Surety A party, usually an insurance company, that assumes responsibility for another party's obligations.

Topographical (topo) survey A drawing prepared by a registered surveyor or engineer showing the boundary lines of a site and the location of buildings, trees, utilities, easements, encroachments, and other information relevant to the site. Usually ordered and paid for by the project owner and included in the bid documents for a proposed project. A variation of a topo survey is a boundary line survey, which may show only the location of the property lines.

References

Bennis, W. (1989). *On Becoming a Leader*, Addison-Wesley Publishing Company Inc.

Carnegie, D. (1981). *How to Win Friends and Influence People*, Revised edition, Simon & Schuster Inc.

Ceridian Corporation. (2006). Ceridian Newsletter Abstracts 13 April 2005. [online] Available from: http://www.ceridian.com/www/content/10/12455/12487/12903/ 12909/041305 _customer_query. html [Accessed 29 January 2006].

Civitello, A. M. Jr. (2000). *Construction Operations Manual of Policies and Procedures*, Third edition, McGraw-Hill.

Corporate Executive Board. (2004). *Driving Performance and Retention Through Employee Engagement.*

Covey, S. R. (1989). *The Seven Habits of Highly Effective People*, Fireside/Simon & Schuster.

Croce, P. (2004). Catching the 5:15. *Fortune Small Business*, March.

Fields, D. (2006). [online]. The benefits of making your banker your friend. Available from: http://www.sba.gov/starting_business/financing/brokerfriend.html [Accessed 22 March 2006].

FMI Corporation. [online]. Bottom-line results. *FMI's 2002–2003 Business Development and Marketing Report in the Construction Industry.* Available from: http://www.fminet.com/global/Articles/MarketingSurvey2003.pdf [Accessed 29 January 2006].

Gitomer, J. (1998). *Customer Satisfaction is Worthless, Customer Loyalty is Priceless*, Bard Press.

The HR Manager (2005). [online]. Corporate culture. Available from: http://www. auxillium.com/culture.shtml [Accessed 15 January 2006].

Knaup, A. E. (2005). Survival and longevity in the Business Employment Dynamics data, *Monthly Labor Review*, May, 50–56.

Knight, B. (2002). *Knight: My Story*, St. Martin's Press.

Lapham, L. H. (1988). *Money and Class in America*, Grove Press.

Maxwell, J. C. (2003). How do you lead in the face of uncertainty? *Atlanta Business Chronicle*, 21 November.

McCullough, D. (2005). *1776*, Simon & Schuster.

Mescon, M. H. and Mescon, T. S. (2001). Excellent leadership is all in the people leading. *Atlanta Business Chronicle*, 7 September.

Powell, C. and Persico, J. (1995). *My American Journey*, Random House.

Rosemond, J. (2006). John Rosemond's Affirmative Parenting, *Parents Newsletter*, ArcaMax Publishing.

Schleifer, T. C. (1990). *Construction Contractors' Survival Guide*, John Wiley & Sons, Inc.

Scott, D. R. (2005). *Apollo Program*, Microsoft Encarta Online Encyclopedia.

Sherman, A. P. (2004). Psst! Pass it on, *Entrepreneur*, March.

Simonson, K. (2005). Quick Facts About the Construction Industry, Association of General Contractors of America.

Smith, Currie & Hancock LLP's Common Sense Construction Law. (2005). T. J. Kelleher, Jr., ed. John Wiley & Sons.

Stanley, T. J. (2001). *The Millionaire Mind*, Andrews McMeel.

Trump, D. J. (2004). *The Way to the Top: The Best Business Advice I Ever Received*, Crown Business.

Welch, J. (2005). *Winning*, Harper Business.

Welch, J. and Byrne, J. A. (2001). *Jack: Straight from the Gut*, Warner Books.

Index